T0216136

Modern Library Chronicles

A SHORT HISTORY

OF MEDICINE

F. González-Crussi

A SHORT HISTORY
OF MEDICINE

A MODERN LIBRARY CHRONICLES BOOK

THE MODERN LIBRARY

NEW YORK

2008 Modern Library Paperback Edition

Published in the United States by Modern Library, an imprint of The Random House Publishing Group, a division of Random House, Inc., New York.

MODERN LIBRARY and the TORCHBEARER Design are registered trademarks of Random House, Inc.

Originally published in hardcover in the United States by Modern Library, an imprint of The Random House Publishing Group, a division of Random House, Inc., in 2007.

LIBRARY OF CONGRESS CATALOGING-IN-PUBLICATION DATA

González-Crussi, F.
A short history of medicine / by F. González-Crussi.
p.   cm.
Includes bibliographical references and index.
ISBN 978-0-8129-7553-6
1. Medicine—History.   I. Title.
1. History of Medicine—Europe.   2. History of Medicine—
United States.   3. History, Modern 1601—Europe.   4. History, Modern
1601—United States. WZ 70 GA1 G559s 2007
R133.G658 2007
610.9—dc22   2006102527

www.modernlibrary.com

CONTENTS

Foreword

It is trite but true that the lessons of history are ambiguous. Nevertheless, the history of medicine offers a perspective from which to verify the stubborn survival, through time and change, of the essential attitudes of medicine: a spirit of inquisitiveness into the origins of disease and a fervent preoccupation with curing or alleviating the suffering disease produces.

The history of medicine has appealed to many authors. Therefore, most of what is told in this book has been told often and ably before. Yet I believe that I can claim some originality, not in the themes developed but in the manner in which they are presented and in the personal interpretation annexed to them. For it is my persuasion that current societal attitudes are out of phase with medical realities and that some light may be shed upon this problem through a look at the historical evolution of medicine.

Today, imaging techniques render the human body uncannily transparent; surgeons operate at a distance by means of robotic devices; organs are replaced using astounding transplant procedures; and functional genes are inserted into cells, thus altering at will the unique fate and individuality that Nature apportions to each living being. In the midst of these scientific and technological marvels, we tend to forget that medicine used to be an art and that, for all its admirable advances, present-day diagnostic and therapeutic methods are still far from possessing mathematical precision.

Most important, today's overwhelming momentum of science and technology tends to make us forget that the foundation of medicine is a universal humanism. My bias is to believe that this fundamental humanistic core is now being threatened and that preserving and safeguarding it is as great a challenge to the profession as any of the toughest technical or scientific problems it has to face.

New epidemics appear (AIDS being perhaps the most obvious example), as do new patterns of morbidity and mortality, which are the inevitable accompaniment of changing world conditions. Then the dark, ancestral fears resurface and enormous pressure is applied to scientists and physicians, who are urged to find quick solutions, forgetting that it is natural for life to be under perpetual risk and that total elimination of disease—though not the alleviation of its burden—is an unrealistic goal.

It is not surprising, when in such a predicament, that people should address their most pressing demands and pleadings to physicians, as these have ever formed part of an influential class. Galen was the physician of Marcus Aurelius; Vesalius, of Emperor Charles V; William Harvey, of the kings of England; and, to abbreviate a long list, in our time we have seen Michael DeBakey looking after Russian president Boris Yeltsin. Physicians' influence is based on the fact that therapy rests largely upon confidence: one must trust the professionals in whose hands one's health and well-being are ultimately deposited. The spectacular progress in their field reinforces physicians' authority, even that of those who had no part in producing the advances. But this exaltation of physicians' image leads some writers to turn the outstanding members of the profession into imposing busts and statues out of a museum. I have tried to avoid this, persuaded that what makes the history of medicine interesting is that it was enacted not by superior and uncommonly gifted people but by men and women precisely like the rest of us, prone to erring and subject to alternating triumphs and disappointments.

Therefore, in the following pages I have not disguised the fact that Louis Pasteur thought nothing of stooping to base tricks of self-promotion unworthy of his name; that John Hunter, the eighteenth-century "father of surgery," was petulant and irascible; that William Stewart Halsted, the American "giant" of modern surgery, became a cocaine and morphine addict; and that Robert Koch, the founder of medical bacteriology, falsely claimed to have found a cure for tuberculosis, and when it was clear that his remedy was useless, he promptly absconded to Egypt, gallivanting around with his new bride while his assistants had to face the consequences of the scandal by themselves.

The protagonists of the history of medicine were men and women like us, but, also like us, they were products of their respective societies. Medical events originate in changes in society and in turn have repercussions upon the latter. Thus, a thorough chronicle of the history of medicine requires reference to the social context of each medical episode. However, such a comprehensive treatment would require greater space than I have at my disposal, plus a learning to which I can make no pretension. I am grateful to my editor, Will Murphy, for authorizing a longer book than was originally assigned. As it is, I found it necessary to lay my subject on a procrustean bed and to excise surgically all that I could not accommodate in the space available. As a result, no chapters devoted to ancient medicine were included, although the continuity between ancient and modern medicine is acknowledged. By the same token, the emphasis is on Western medicine since the inception of the scientific method, but the contributions of the Orient, and of epochs predating the dominance of the rational spirit, are not ignored.

Despite the many deficiencies that will undoubtedly be found in this book, I will consider my task accomplished if I can transmit a certain sense of regard for medicine. That its history does honor to it should be beyond question, since for all the bunglings and failings— sometimes even crimes—of its practitioners, it remains an essentially altruistic endeavor. And in a world that often seems bent on using science and technology to create better means of mass extermination, one can only think well of a discipline whose goal is to devise better ways to alleviate the suffering and to cure the ills of our fellow men.

1

THE RISE
OF ANATOMY

If Western medicine is unique, it is because it made the body an object of systematic, scientific study. This is not stating the obvious. The human body has innumerable symbolic meanings, all emotionally charged and often contradictory. Turning it into an object of orderly inquiry and meticulous investigation was no small achievement.

It seems that for some cultures, the body hardly exists at all. Certain aborigines of New Caledonia, in the South Pacific, use the same words to name the parts of the body and the plants or other objects of their natural environment, between which they perceive a resemblance. For instance, the skin of the body and the bark of trees are designated by the same term; the identical word is used for the flesh of human limbs and the pulp of fruits; and the various inner organs share names with the produce that they outwardly resemble. In this society the body is not thought of as an independent entity but is indistinguishable from its surroundings.[1] Similarly, in European societies during the Middle Ages and in certain communities in more recent times, alchemical notions have linked various parts of the body to the constellations of the sky: Aries "rules" the head; Leo, the heart; Scorpio, the genitals; and so on. The body lacks a clear border; in the elemental imagination, it merges with the rest of the cosmos. To cut or incise the body, as in anatomical dissection, would have seemed an aggression against the continuum that linked man and his environment, an attempt against the unity of the world. The impulse to study the body's anatomy could hardly have arisen in a society in which such concepts prevailed.

Other societies surrounded the body with religious sentiments. In the Judeo-Christian tradition, in which man was created in God's image, the body is a temple. Thus, it deserves solemn respect. On the other hand, the same tradition gave rise to ascetic movements in which the body was a repository of sin, a despicable, filthy thing that ought not to be made the center of an honest man's concerns, much less an object of serious study. Either stance was contrary to anatomical investigation. In the Middle Ages, all intellectual activity resided in the Catholic Church, but no clergyman ever distinguished himself as a surgeon or even as a barber (the profession which was then in charge

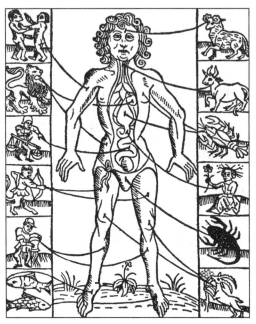

Correspondence between the zodiac signs and the parts of the human body, according to a German medieval illustration. From Martyrologium der Heiligen nach dem Kalender *(Strasburgh, 1484).*

of minor surgical procedures), due to the severe ecclesiastical prohibition against all forms of bloodshed.

For all their admirable medical insights, ancient India and China did not make the interior of the body the basis of their medical systems. And neither the ancient Egyptians nor the ancient Mexicans contributed anything of substance to the knowledge of anatomy. It is astonishing that the opening of innumerable human bodies in ritual practices—the Egyptians embalmed tens of thousands of human and animal cadavers; the Aztecs sacrificed countless victims by opening the chest and extracting the heart—should not have sparked curiosity about the structure of the organs they exposed. Yet their attitude was different: they looked at the world from a mythicoreligious perspective that was incompatible with an impersonal regard of objective reality. Thus, conceiving of the body as an autonomous object and deeming it worthy of study—these are the signal achievements of Western medicine.

The Greeks, as is often the case in the history of Western civilization, take the palm for intellectual curiosity. Still, Hippocrates (c. 460–c. 377 B.C.) knew very little anatomy and did not seem interested in correcting his ignorance: there is no record in the entire Corpus Hippocraticum—his own writings and those of his followers—of any mention of anatomical dissection. Aristotle (384–322 B.C.) constructed a remarkable system of anatomical knowledge, all the more admirable when one reflects that he never dissected a single human body. It was entirely based on dissections of animals: birds, reptiles, mammals, and especially monkeys. But Aristotle was above all a thinker: the body was for him principally an object of metaphysical speculation.

Aristotle was a student of Plato, for whom the world of the Ideal took precedence over the more pedestrian here and now. The Aristotelian philosophical system is formidable, but when it deals with corporeal form it becomes sketchy. Aristotle, the famous Stagirite, seeks to explain the human body's position in the universe, how it came into being, what its origins are, and the meaning of its life. The details of bodily structure are secondary to the comprehensive nature of his metaphysics.

Two scholars of the Hellenistic civilization did perform anatomical dissections of human bodies: Herophilus (c. 335–c. 280 B.C.) and Erasistratus of Ceos (c. 325–250 B.C.). Both settled in Alexandria, Egypt, a great intellectual center of the ancient world, where Greek communities had long existed and which was the site of the fabled library containing more than half a million volumes, eventually destroyed in a fire. Little is known of these two men.[2] They left no written works and are known only from references by other authors.

Herophilus was born in Chalcedon, an ancient town situated near today's Istanbul, Turkey. He is credited with having named the duodenum (so called because it is twelve—_duodeni_—finger breadths long) and the prostate (Greek _prostates,_ "standing before," since it is placed before the rectum); and for having determined that the arteries are full of blood, not air, as was commonly believed. This misconception may have arisen because arteries, having thick musculoelastic walls, tend to contract postmortem, so that in the cadaver they are usually empty; the blood, having been squeezed out of them, fills the veins, which have thinner, distensible walls. Herophilus also described several brain structures, including a site of cranial venous confluence that still bears his name (*torcular Herophili,* or "Herophilus's press"). He traced the course of nerves to their origin in the brain, thereby firmly establishing that impulses for voluntary movements travel from the brain, the site of the will and the reasoning faculty, to the extremities via the nerves, not via the arteries, as was wrongly believed.

His younger colleague and collaborator Erasistratus was born in a hamlet on the island of Ceos (or Keos) and studied in Athens. He described the cardiac valves, named the tricuspid valve, and confirmed and extended many of Herophilus's observations on the cranial nerves. He stuck to the belief that arteries carry air and explained the bleeding that follows their severance by proposing that the arterial walls, like all tissues, are made of tightly woven tiny veins, which promptly bleed to fill the vacuum that ensues when the artery is cut. The explanation may strike us as far-fetched, but it was perfectly in keeping with the concepts of his time. And it took genius to realize, long before the invention of the microscope, that all tissues possess innumerable small blood vessels (today we call them capillaries), aggregated into a dense network. Erasistratus imagined that the nutriment carried by these vessels poured between the spaces of the net, the *parenchyma* (Greek for "something poured in beside").

His realization that the cardiac valves function like guards that impede retrograde flow was no less astounding, given that it came eighteen centuries before the discovery of the circulation by William Harvey (in 1628) and at a time when no valve-based propelling pump had been invented. He may be excused if his explanation of the one-way valves is part of a theoretical system that today sounds like pure nonsense.

There is a somber note that dims the glory of these two outstanding savants. Apparently, the Aristotelian idea that true knowledge of bodily structure is possible only by studying a living being prompted them to vivisect men. The kings of Alexandria, desirous of maintaining their city as a leading center of the arts and sciences, granted permission for condemned criminals to be officially surrendered to the anatomists, who then "legally" laid them open.

Some historians have cast doubt on the reality of that practice. If it did take place, one shudders to think what scenes of indescribable torture may have taken place in the dissection rooms of Herophilus and Erasistratus. The miserable, wretched victims were slowly cut open, their bloody, trembling organs exposed, turned over, palpated, and inspected; all amid shrieks of pain and under the cool glance of the anatomists and their pupils and assistants. Little wonder that such Christian writers as Saint Augustine and Tertullian fulminated against them and their practices, calling them ferocious beasts and bloody butchers. Again, one must place their actions in the proper historical context. The Roman Empire was on the rise. These were times when weekend family entertainment consisted of watching gladiators hack each other to death, wild animals devour human beings, and other similar "amusements." Some spectators of these "games" jumped into the arena, rushed over to an agonized gladiator, and drank his fresh blood or tore out a piece of his warm liver to eat, in the belief that such ingestion could cure epilepsy.

After the brilliant Alexandrian period, the study of anatomy waned into the intellectual lethargy of the Middle Ages. Interest in human anatomy was briefly rekindled by the famous Galen (A.D. 129–c. 199), who for some time held the post of physician in charge of the gladiators of the arena of his native Pergamum (today Bergama in Anatolia, present-day Turkey). In this capacity, he observed the horrible mutilations and ghastly tears that the fighters sustained, and through the gashes he certainly observed, as best he could, the conformation of the internal structures. But public opinion had changed: there was now great respect for the cadaver, and sentiment was strongly against perturbing the soul's peace in the hereafter by cutting up the body it had once tenanted. Anatomical dissection was viewed as a profanation. Ironically, this culture, which was so disrespectful toward living humans—as seen in the oppressive treatment of their women, the cruel abuse of their slaves, and the merciless, bloodthirsty character of their amusements—held the dead in great reverence.

Galen displayed prodigious activity. He dissected thousands of animals in public demonstrations: reptiles, birds, camels, bears, dogs, weasels, rats, mice, lynxes, and even elephants. But he favored pigs and monkeys for his demonstrations, because of the purported similarity of their organs to those of humans. Unable to assuage his curiosity, he did not waste any opportunity to look at human bodies whose insides had been fortuitously exposed. Thus, he once stopped by a town that had been ravaged by a flood, and while everyone else was running about loudly lamenting the disaster, he was delighted to see that a corpse, tossed ashore by the receding waters, had decomposed in such a way that various parts were still articulated and freed from concealing tissues, "as if it had been prepared by an expert anatomist."[3]

The anatomical concepts owed to Galen's indefatigable labors became dogma. They were based largely on extrapolations from findings in animals, but they would last unchallenged for a thousand years. Galen himself, in spite of his great intellectual curiosity, felt that there was a knowledge of anatomy that was "necessary" and another kind that was "superfluous." Physicians had very few treatment options at their disposal. They could reduce bone dislocations, set fractures, and perform certain procedures that required no more than a cursory knowledge of the muscles and joints of the limbs. They developed treatments for the injuries of the gymnasium or the battlefield. But attempting to intervene inside the body was off limits: intrathoracic or intra-abdominal operations were almost invariably fatal. Hence, Greek physicians adhered to the opinion of their most prominent brethren: that a physician should have some knowledge of anatomy, chiefly of the extremities, but to worry about the cardiac valves or the cerebral ventricles was foolishness. Would they be able to treat diseases of these structures? Their time would be better invested in learning what was possible.

Medicine in the Roman Empire, having achieved its climax with Galen, went into a state of decline, despite the efforts of some isolated physicians of the Byzantine historical period, such as Oribasius of Pergamum (c. 325–c. 400), Aëtius of Amida (502–575), and Paul of Aegina (Latin, Paulus Aegineta: c. 625–c. 690). Scientific knowledge was, in fact, kept alive by Arabic scholars. The world owes an incalculable debt to the scholars of what has been variously called Islamic, Arab, or Arabic medicine (although they were not all votaries of Islam nor necessarily ethnic Arabs) for having perpetuated the classic legacy of Greece and Rome.[4] These were men such as Avicenna (Abu Ali al-Husayn ibn Abd Allah ibn Sina, 980–1037), Rhazes (Abu Bakr Muhammad ibn Zakariya al-Razi, c. 865–c. 925), Maimonides (Moshe ben Maimon, 1135–1204), Albucasis (Abu al-Qasim, c. 936–c. 1013), and others. (These physicians will be discussed at greater length in the next chapter.)

Since Arabic, the language of the Qu'ran, was the lingua franca during the expansion of Islam, these scholars were able to establish links with the traditions of the Far East, including India and China, a fact that tends to be overlooked by medical historians. Although remote contacts are obscured by fictional tales and legends spawned by the proverbially fertile Oriental imagination, there is firm historical evidence of close intellectual cooperation between China and the Arab world from at least the seventh century well into the eleventh. A great Islamic physician of the early fourteenth century, Rashid al-Din al-Hamdani, wrote a book that contained much Chinese traditional medicine and even proposed that the Chinese written language should be adopted as the language of science, on account of its being less ambiguous. Conversely, one Chinese physician, a member of a group of scholars who visited Rhazes, stayed with this famous alchemist-physician for a year, learned to speak Arabic, and announced his intention to copy the sixteen books of Galen, which were at the time the basis of Islamic medicine. A tradition says that the Chinese physician, whose name is forgotten, not only copied the voluminous texts but amazed everyone by copying faster than it was dictated to him by using a shorthand method of his own invention. But "something must have happened on the way back to China," says the great orientalist Joseph Needham regretfully, because there is no trace of Galenic medicine in the Chinese medical body of knowledge.[5]

In spite of the valuable contributions of Arabic medicine and the cross-fertilization that it fostered, Islam prohibited anatomical dissection, and the study of bodily structure did not advance significantly. Thus there was a prolonged hiatus from anatomical study during Roman times up to the Renaissance.

According to evidence, human dissection was conducted again as early as 1315 in Bologna, Italy. Mondino de' Liuzzi's textbook *Anatomia mundini*, written in 1316, became the standard text on the subject. Though undeniably important for renewing interest in anatomy, the book was no marvel. It was a Galenic text, therefore perpetuating the many inaccuracies and conceptual errors that the master of Pergamum had turned into articles of faith by the sheer weight of his authority. Moreover, it was a Latin translation from the Arabic, and a mediocre one at that. But if its immediate impact on medical education was limited, its importance to the long-term development of anatomical science cannot be underestimated. Among other things, it emphasized the importance of dissection as a superior way to gain firm knowledge.

The engravings of medieval anatomy texts help us to understand how the dissections were conducted. These illustrations generally

Scene of a medieval lesson of anatomy, redrawn from John of Ketham's Fasciculus Medicinae *(1493), itself after an illustration from a text by Mondino de' Liuzzi (1316). The prosector and the demonstrator work on the cadaver; the lector sits on an elevated platform; and other robed academics are in the background in a precisely determined order.* COURTESY OF THE NATIONAL LIBRARY OF MEDICINE

show the cadaver lying on the table, surrounded by men participating in the learning experience. Under the table there is a basket or some other receptacle used to collect the debris from the dissection. Holding a knife and about to use it on the cadaver is the *sector* (today's term is "prosector"), i.e., the individual in charge of actually performing the dissection; he is easy to identify because he is the only person not attired in academic raiment. This man is probably a barber or a surgeon, occupations not held in high esteem at that time; in any case, the *prosector* was a man of limited instruction.

The illustrations usually show another man by the table: the *ostensor,* or demonstrator, identifiable by a stick, which he uses as a pointer. Since the prosector was an ignorant man and could not understand Latin (the language in which the lessons took place), the *ostensor* indicated to him where to cut and what to do next.

Behind them, receded from the table and raised high above it, we see the *lector,* or "reader." He is the professor of anatomy, who sits on his chair, behind a lectern, and reads a text from Galen's works or recites it from memory. The others congregated around the table are students or instructors. The professor's solemn and dignified air, his removal from the disagreeable sensations of the dissecting table, the high position of his chair, and the fine academic attire in which he is clothed all seem to indicate that his was the highest rank in the anatomists' milieu. On the other hand, some historians contend that the most knowledgeable and experienced person was the demonstrator, who was close to the corpse and had firsthand knowledge of the structures he demonstrated. Apparently, some professors acted alternately as readers and demonstrators.

However, in the midst of all this paraphernalia ostensibly aimed at teaching anatomy, the fact remains that the real aim was quite different. Historians and scholars have not reflected enough on this extraordinary situation: the professor read, the students listened, the demonstrator pointed with his stick, and the prosector cut and rummaged in the cadaver's innards. But no one learned: the pragmatic end of medical and anatomical teaching was never realized during the Middle Ages. As the historian Nancy Siraisi[6] has pointed out, in reality this experience was intended only to elucidate the ancient texts!

This was the paradoxical situation when the powerful figure of Vesalius entered the world stage. A native of Brussels, Andreas Vesalius (1514–1564) came from a family celebrated for the number of physicians it produced. His great-grandfather had been a court physician to Mary of Burgundy, the wife of Maximilian I; his grandfather wrote a book of commentaries on the *Aphorisms* of Hippocrates; and his father was apothecary to Charles V before the latter became emperor.

Vesalius was a man of unflinching determination and fixity of purpose. He studied in Louvain, Cologne, and Montpellier. In 1535 he

Portrait of Andreas Vesalius, from his book De humani corporis fabrica, libri septem *(1543).*

went to Paris, where he performed numerous public dissections, and then to the Low Countries and to Italy. In later years he recalled the difficulties and privations he had to undergo in order to fulfill his dream—one might say his obsession—of acquiring an exact knowledge of the structure of the human body. As recounted by Vesalius, he spent long hours in the Cemetery of the Innocents in Paris, turning over bones. Once, with a companion, he was attacked by a pack of savage dogs; he was locked out of the University of Louvain when he went, in the middle of the night, to take away from the gibbet the bones of an executed man, in order to prepare a skeleton; he kept decomposing cadavers in his bedroom for several weeks, suffering the stench; and he urged medical students to take notice of the most gravely ill patients, that they might later seize their corpses.[7]

A man of this mettle would not have been expected to conform easily to received opinion. His firsthand experience with dissection told him that the Galenic texts were riddled with inaccuracies and unwarranted extrapolations of anatomical features from animals to man. He realized the deficiencies and dared to challenge openly what had been dogma for a thousand years. His impressive knowledge, boldness, and ambition earned him both enemies and followers. In 1543 he was appointed physician to the court of Emperor Charles V, and when the latter abdicated, he remained as court physician to Philip II, king of Spain.

His biography reads like a novel. While at court, his fortunes rose after successfully treating a member of the royal family. The Infante Don Carlos, the ill-fated son of King Philip (whose name today is remembered chiefly thanks to Giuseppe Verdi's opera *Don Carlos*), sustained an accidental fall—apparently, he tripped down a staircase while in pursuit of a pretty maid. His head wound became infected, and his general state of health was seriously compromised. The most eminent physicians of the region were summoned, and Philip sent his own personal *protomédico,* but their ministrations had no effect. Vesalius was regarded with distrust: after all, he was a native of Flanders, then in open rebellion against the imperial power of Spain. And his fame as a physician did not ingratiate him to his envious colleagues in the Spanish royal court. His advice was ignored.

As the prince's health deteriorated, however, the general consternation deepened. The patient's appearance was appalling; the swelling spread relentlessly, distorting his face, while an unremitting fever weakened his body and upset his mind. The concerned medicos became alarmed, and when their noble patient sank into a coma they grew desperate. Appeals were made to religion where science had failed. A procession of hooded flagellants paraded in front of the prince's sickroom, furiously hitting their bare backs with whips, straps, cat-o'-nine-tails, and other instruments of torture. Yet this caused no improvement in the patient's condition. Then, the mummified body of a Franciscan monk, Brother Diego de Alcalá, who had died in odor of sanctity nearly a century before and was widely reputed to possess miraculous powers, was laid overnight in the same bed as the languishing prince. To the court's chagrin, even this proximity to the sacred relic did not ameliorate the Infante Don Carlos's clinical course.

As a last resort, the prince's caretakers, fanatic Catholics though they were, had no compunction in letting a Moorish quack from Valencia, a Muslim popularly called Pinterete, prescribe some nostrums. When these did not cure the young man's affliction, only then did they finally accept Vesalius's proposed therapy, which included a trepanation by incising the outer orbital wall. This was done, and Don Carlos made a fast recovery. As a result, Vesalius's renown skyrocketed, as did the esteem that King Philip had for him. Yet the king attributed the cure to the night that the infante had spent beside the remains of blessed Diego de Alcalá and not to the bold treatment by Vesalius.

Vesalius's death was as dramatic as his life. He undertook a pilgrimage to the Holy Land, under circumstances that are not quite clear. One version has it that he started the dissection of the body of a recently deceased and highly regarded person and was shocked to see, upon opening the chest, that the heart was still beating. In order to escape arrest by the Inquisition, Vesalius relied on the royal protection he enjoyed and sailed to the Holy Land in atonement for his fault.

Be that as it may, he never came back. In one account, the ship he had boarded went down off the coast of Zacynthus, one of the biggest Ionic islands of Greece and the most southern. The great anatomist managed to reach the shore but died from his injuries. Less exciting, but probably more credible, is the version that says that he became ill on board and had to be disembarked because the crew suspected the plague and was afraid of contagion. Regardless of which narrative is true, Vesalius lies buried amid the sunny vineyards of Zacynthus.

Vesalius's enduring fame rests largely on the book he authored, *De humani corporis fabrica, libri septem* ("The seven books of the structure—*fabrica*—of the human body"). It is a superb work. William Osler (1849–1919) saluted it as the greatest medical book ever to appear, and medical historians ranked it number one among the "ten greatest medical discoveries," together with the discovery of the circulation by William Harvey.[8] Insofar as the book accurately describes and magnificently depicts the body's component parts, this praise is no exaggeration. The multisecular nonsense of the Middle Ages was finally toppled. Physiology could now be constructed on a solid base.

The book demolished fantastic notions such as the existence of pores interlinking the cardiac ventricles, the division of the uterine cavity into seven chambers, the direct links between brain and testicles, and so on. Still, the impact of Vesalius's book might have been more muted had it not been for its supreme elegance and artistic value. The identities of the draftsmen who produced the woodcuts used to illustrate the *Fabrica* are disputed, but it is known that they worked in the shop of Titian (Tiziano Vecellio, 1488/90–1576), the great Renaissance master, and it has been suggested that Titian himself had a hand in the design or execution of at least some of the drawings. Vesalius supervised the work zealously, ensuring that no inaccuracies found their way into its pages. Every detail had to be checked and rechecked against the real models. Vesalius had performed the dissections himself, disregarding the authority of Galenic pronouncements, and was unwavering in his demands that the artists adhere as closely as possible to what they saw and ignore the suggestions of their aesthetic sense.

The figures are anatomic illustrations as much as they are splendid and somewhat troubling pieces of art. Some of these remain famous to this day. To demonstrate the muscles, we are presented with men whose skin has been peeled off. Flayed human beings, either in two-dimensional representations or in statuary, were widely used during the Renaissance for anatomical instruction. The theme of the flayed men (French: *écorchés;* Italian: *scorticatti;* Spanish: *despellejados*) had special resonance in that era, which was enamored of Greco-Roman antiquity, because it echoed the story of Marsyas, a satyr who was flayed by Apollo in punishment for having dared to defy the god in a flute-playing contest. The myth was familiar to all educated persons in the Renaissance.

The men's arm and leg muscles have been detached from their upper insertions and hang pendulous, in tatters. Their body cavities are exposed, their crania opened. In spite of all this, they adopt elegant poses, as if dancing a minuet, reading a book, or taking a stroll in an idyllic landscape. These are cadavers that adamantly "refuse to play the role of corpses," as a French art critic and essayist once put it.[9] The blatant contradiction between harrowing bodily devastation and nonchalant attitude creates a dreamlike effect.

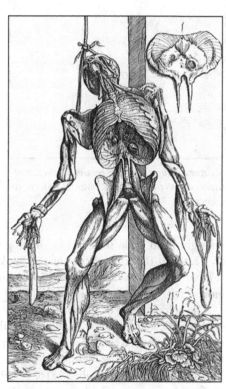

In one of the engravings, a flayed man whose abdominal wall has been removed is hanging from a rope. Yet the rope is not tense but slack, and, as a result, the corpse is not fully erect: its knees are flexed, and its head is tilted backward. The figure's arms are partly outspread, the muscles of its forearm hang loose from their lower insertion on the fingers, like sleeves slashed lengthwise, and the long strips still cling to the cuffs. It is a powerful image. But the man's diaphragm, nailed to a wall, reminds us that the

A strangely evocative illustration from Vesalius's book.

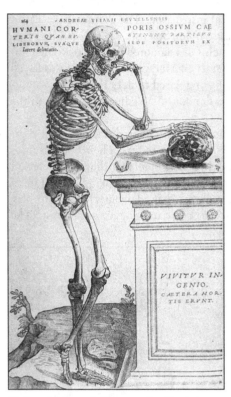

Skeleton holding a skull in a pensive, Hamletian attitude. COURTESY OF THE NATIONAL LIBRARY OF MEDICINE

main goal of the illustration is strictly pedagogic. Another figure still has some of his facial skin, conveying a strange, moving sense of sadness. He is also there for our instruction, though: we see that he has carefully deposited on the ground, against a wall, a part of his skeleton.

Vesalius was especially concerned with osteology. His book devotes the first and greatest space to the study of the skeleton. Some of the most splendid illustrations are found here. One of the best known shows a skeleton standing in a pensive attitude, his left elbow resting on a high table and the dorsum of his left hand against his face. This skeleton "gazes" at a skull that rests on the table and delicately poses his right hand upon it. It is a fascinating image of death meditating on death. In the first edition of the book, the table was captioned, *"Vivitur ingenio, cœtera mortis erunt"* ("Genius lives, everything else is mortal").

The achievement of Vesalius was truly revolutionary: the concept of the body, and hence of medicine, would never be the same. This work was continued and expanded upon by his disciples, only the most famous of which can be mentioned here.

Matteo Realdo Colombo (1516?–1559) succeeded the master as professor of anatomy and surgery at Pisa, Padua (1545–1548), and later Rome (1549 to his death). As is not uncommon, friction arose between master and pupil. Colombo accused Vesalius of inconsistencies and erroneous descriptions; notably, that he had used the eye and tongue of an ox, not of a human being, in his published presentations. Colombo's chief work was entitled *De re anatomica libri XV,* posthumously published. He was a personal friend of Michelangelo and wanted to recruit him as illustrator for his book. The age and infirmities of the great artist did not permit the realization of this project, which, had it been done, might have outshone even the *Fabrica.*

The fifteenth "book" (today we would say "chapter") of *De re anatomica,* which is the most extensive, initiated the study of pathological anatomy. The anatomist describes not normal structures but a collection of abnormalities that he had encountered in the course of his investigations,[10] including intracardiac thrombi, venous calcifications, renal and hepatic calculi, esophageal varices, imperforate hymens, and so on. It fell to Colombo to perform the autopsy of Saint Ignatius of Loyola, founder of the Society of Jesus, the religious order of the Jesuits. The saint was afflicted with numerous biliary calculi and perhaps "stones" in other organs (the text of the report is ambiguous).[11]

Colombo may have discovered the pulmonary circulation, although this is hotly disputed. Some historians believe the Arabic doctor Ibn al-Nafis (d. 1288) had made this find almost three centuries before. In addition, Miguel Serveto (1511?–1553), the Spanish theologian-physician executed for heresy by the Calvinists of Geneva, had mentioned his discovery of the pulmonary circulation in his book entitled *Christianismi restitutio,* a mostly theological work.

Bartolomeo Eustachio (1520–1574), another successor of Vesalius, in some respects outdid the master. He focused on oral anatomy and described in great detail the morphology and development of the teeth. He was the first to identify the dental pulp, the root canal, enamel, dentin, and the periodontal membrane; he also examined the musculature of the mouth and the tongue and may thus be thought of as the founder of modern dentistry. His investigations also extended to other regions of the body. Some consider him a pioneer in studies of renal anatomy,[12] and students of anatomy recognize his name in the "eustachian" or auditory tube, a conduct that connects the tympanic chamber with the nasopharynx, thus allowing an equalization of pressure on both sides of the tympanic membrane.

Gabriel Fallopius (1523–1562), one of Vesalius's pupils, wrote *Observationes anatomicae,* which includes descriptions of the anatomy of the middle and inner ear, with masterful dissections of complex structures such as the cochlea and the ossicles of the ear. He was interested in the reproductive apparatus of both sexes, and his name became attached to the oviducts, or *fallopian* tubes. Curiously, his name is also associated with the history of the condom. Having heard that syphilis had been brought by the early explorers from America to Europe, where it became particularly virulent, he set about to design a penile sheath. It was made of linen—rubber and latex had not been invented—apparently cut to individual measure, imbued with herbs and salts for therapeutic purposes, and tied with a colorful ribbon at the base, for the sake of aesthetics. In his book *De morbu gallicum* ("On the French disease"), he said he had conducted a clinical trial on more than 1,100 men, none of whom contracted syphilis. However, it is a mistake to credit Fallopius as the "inventor" of the condom. Such sheaths had been in use from remotest antiquity—made of reed in ancient Egypt, of oiled silk in ancient China, and of fish tripe or animal intestine in various countries at various times.

Hieronymus Fabricius ab Aquapendente (1537–1619), the fa-

Fabricius ab Aquapendente (1537–1619).
COURTESY OF THE NATIONAL LIBRARY
OF MEDICINE

vorite pupil of Fallopius, has been called the "preceptor of [William] Harvey." His studies of the circulatory system, in which he gave special attention to the valves of the veins,[13] suggested to him the centripetal direction of the circulating venous blood (hitherto believed to be centrifugal). His study of the development of the chick embryo was so exhaustive, and the illustrations annexed to it so elegant, that some consider Aquapendente the "father of modern embryology." His embryologic ideas were developed in *De formato foetu* (Venice, 1604) and other treatises, a modern English translation of which is available.[14] Interestingly, he devoted one of his works to the language of animals (*De brutorum loquelâ;* Padua, 1603).

Although Fabricius ab Aquapendente has been called "the last of the great anatomists of the Vesalian era," he was neither the last nor a direct disciple of Vesalius. The sixteenth and seventeenth centuries were the golden era of descriptive anatomy. Numerous scholars were devoted to the ascertainment of bodily structure. In Italy, these included Leonardo Botallo (1530–1571), Cesare Aranzio (1530–1589), Giulio Casserio (1552?–1616), and others. Elsewhere, eminent investigators appeared, such as the German Johann Georg Wirsung (1589–1643; "Wirsung's duct" is the excretory duct of the pancreas) and the Dutch Adriaan van den Spieghel (1578–1625; "Spieghel's lobe" is the quadrate lobe of the liver), and such luminaries became more and more numerous as time went by.

The race to map the entire human body was on. As navigators left their names on the territories they "discovered," so anatomists, cartographers of inner space, became eponymic designations for the newly revealed body parts. But they were all the spiritual progeny of Vesalius, for they inherited the intimate conviction that anatomical knowledge was to be plucked out with their own hands, not those of a menial; and that credence was to be given to their own eyes, not to the word of the ancients or the power of authority.

2

THE RISE
OF SURGERY

Surgery is the "rags-to-riches" story of medical disciplines. Surgeons rose from despised, uneducated empirics to adulated superstars of vanguard medicine, from callous dabblers in vivisection to sophisticated professionals, aided by the most advanced resources that science and technology have to offer.

The origins of the trade go back to remotest antiquity. There is evidence that trepanation, the making of a hole in the skull for magico-religious reasons or for the treatment of traumas and various diseases, has been practiced since prehistoric times. This was especially so in Peru, where the *sirkaks* (Incan surgeons) reached such a high degree of skill that they treated cranial fractures, epilepsy, infection, scalp diseases, mental illness, and other conditions by trephining. Their trephines were done by drilling, circular-cutting, crosscut-sawing, or scraping the skull. They used gold, silver, or bronze chisels, obsidian knives, curettes made of whale's teeth, and other primitive instruments that served as bone elevators, protectors for the meninges, suturing materials, tourniquets, and so on. With these resources, and in precarious circumstances, these native Americans performed the operation with a 70 percent survival rate,[1] as evidenced by bone-healing around surgical holes in skulls. In contrast, in eighteenth-century Europe, trepanation carried a nearly 100 percent mortality rate. In the first half of the nineteenth century, surgeons joked that anyone who dared to open the skull of a patient must have previously sustained a fall on his own head!

It is thought that the Peruvians made use of the antiseptic properties of tree resins, such as so-called Peru balsam (a resinous substance that flows from a leguminous tree, *Myroxylon pereirae*), in addition to tannin, saponins, and cinnamic acid, substances also used in the embalming of cadavers. For anesthesia, the likely agents came from herbs that were native to their Andean habitat, such as coca, yucca, and others; alcoholic beverages made from fermented corn (*chicha*) may have served a similar purpose. As a further defense against infection, the ancient Peruvians did not live, as we do, in the crowded conditions of enormous urban settlements. Crowding, by permitting multiple successive passages of bacteria from one person to another, increases the virulence of microorganisms. In the transfer of bacteria from one culture medium to another, successive generations are reinvigorated. Additionally, the ancient Peruvians did not have our hospitals. It is no mystery that today's hospital-acquired ("nosocomial") infections are a major cause of morbidity and mortality.

So impressed were twentieth-century Peruvian neurosurgeons with the results obtained by their ancestors that they decided, in a bold and questionable move, to operate on a couple of patients using the ancient Incan trepanation instruments from the national archeological museum. The published results of this odd attempt at "experimental history" show that all the phases of the operation were remarkably well served by the ancient instruments.[2]

For evidence of other early contributions to the surgical art, we must look to India, the country Mark Twain called "the mother of history, the grandmother of legend, and the great grandmother of tradition." Here arose the system of medicine that in the post-Vedic period (800 B.C. to 1000 A.D.) came to be known as Ayurveda ("science of life"), most of whose formulas were contained in texts that repeated the hoariest oral traditions. The text known as *Sushruta samhita* deals mainly with surgery; it contains 184 chapters and mentions 1,120 different pathological conditions.

Although the *Sushruta samhita* enjoined the study of anatomy, how was this to be done in a culture that abhorred the direct cutting or handling of a cadaver? The cadaver was to be placed in a cage and submerged into a hidden spot of a river. There, the body would be allowed to decompose, and after several days the would-be anatomist could sweep off the crumbly tissues with a whisk made of grass roots. This way, there would be no direct contact with the cadaver, and the student could examine "every organ, great or small, internal or external." Clearly, the *Sushruta* method of peeling off rotting tissues was neither practical nor fit for demonstration of structural details; it is doubtful that many students learned much anatomy this way.

The *Sushruta* recommended a number of exercises to aspiring surgeons: incising certain fruits with a sharp knife; suturing pieces of cloth together; bandaging stuffed dolls; probing the inside of bamboo reeds; extracting the seeds of certain fruits; and so on—all this to steady the pulse and to develop manual dexterity. The practiced surgeon would thus become adroit in the use of his instruments, an admirable array of which had been developed. Interestingly, forceps, probes, scissors, knives, and sundry other instruments were often fashioned in the shape of animals. We can picture the ancient Indian surgeon extracting a thorn from a patient's foot with a "raven's beak forceps" or asking his assistant for the "lion's head" when trying to grasp a bulkier foreign body.

Suturing was done with silk from China, as well as hemp, linen, cotton, and plaited horsehair (the Peruvians sometimes closed scalp wounds by simply tying the hair so as to hold together the lips of the wound). Heads of ants doubled as staples, an ingenious use of a natural resource found not only in ancient India but in some preliterate cultures until very recent times. The mandibles of certain ants, which are curved like fishhooks and extremely sturdy, make excellent staples to secure the edges of a wound, as elegantly demonstrated and illustrated with photographs by Dr. Guido Majno.[3]

With these resources the Indian doctors did wonders. For the repair of torn earlobes, their techniques predated those employed by European surgeons by at least two thousand years. Likewise, the repair of mutilated noses (rhinoplasty) devised in ancient India—lifting a flap of skin from the forehead and bringing it down to fashion a new nose—has remained essentially the same to this day.

Their knowledge of anatomy may have been sadly deficient, but centuries of practical experience had taught them important facts. They recognized certain points of the body, which they called *marma*s, as being endowed with special characteristics and knew which ones, if injured, were likely to have serious consequences. Thus, they may not have known anything about the nerves, but they were aware that certain *marma*s, if seriously hit or penetrated, would produce paralysis. Likewise, they could judge which points of the body were likeliest to produce serious hemorrhages and which would cause death if a penetrating weapon were suddenly withdrawn.

But for all their inventiveness and ingenuity, the surgeons of the ancient world, in India and elsewhere, had to limit their ministrations largely to the exterior of the body and the extremities. To open the chest and abdomen was tantamount to killing the patient. This was true for several reasons: ignorance of anatomy; the ever-present risk of exsanguination; the unbearable, shock-producing pain attending major surgery; and the infections that followed the surgical act. The surgical profession would rise against this four-headed hydra. But its progress would not be linear and unimpeded; it would be irregular, zigzagging, and sometimes punctuated by exasperating periods of apparent retrogression.

The rich tradition of antiquity, inherited from Egypt, Babylon, and later Greece and Rome, was passed on in the Middle Ages to the brilliant Arabic doctors. The last great physicians of the Greco-Roman culture had already begun to advance the surgical art. Oribasius (c. 325–c. 400), a physician born in Pergamum, like Galen (whose work he compiled in seventy volumes of his *Collectiones medicae*)[4], excised venereal condylomas and recognized cancerous lesions incurable by surgery. Paulus of Aegina, or Aegineta (625–690), wrote an *Epitome* in seven books, the sixth of which was devoted to surgery. In it he offers instructions on how to arrest hemorrhages, descriptions of techniques for excision of nasal polyps, circumcision, hemorrhoidectomy, treatment of varicose veins, and methods of extracting arrows from various parts of the body, as well as the first detailed description of tracheotomy.[5]

The famous Persian doctor Rhazes (Abu Bakr Muhammad ibn Zakariya al-Razi, c. 865–c. 925) made some isolated attempts at performing abdominal surgery and is said to have been the first to use animal gut sutures. According to tradition, the caliph consulted him on the best place to construct a hospital in Baghdad. Rhazes ordered pieces of meat to be hung in many sites of the city and chose the one in which the meat took the longest to spoil (presumably meaning that there were the fewest flies present). Rhazes authored an encyclopedic work, the *Kittab al-hami,* which gathered practically all the medical knowledge available in the tenth century, and his *Treatise on Smallpox and Measles* became a respected reference work for several centuries.

Avicenna (Abu Ali al-Husayn ibn Abd Allah ibn Sina: 980–1037), perhaps the most influential of the healers and philosophers of the Islamic world, composed a medical work that became known in the West as the *Canon of Medicine.* In philosophy, his ideas are said to continue their influence in certain currents of Islamic thought to this very day. However, his contributions to surgery were limited. Nonetheless, Avicenna was a remarkable surgeon and was apparently the first to use oral-tracheal intubation in conditions of impending asphyxia.

Albucasis (Abu al-Qasim, c. 936–c. 1013), a Hispanic-Arabic physician (he was born near Cordoba, Spain), achieved great notoriety in Europe. His book *Al-Tasrif* was a multivolume medical encyclopedia with an important section on surgery. Like most Arabic surgeons, he resorted to cauterization of large vessels as a means of stanching bleeding. He described the extraction of a bladder stone (a condition apparently very common in the Middle Ages and the early modern era) by inserting a finger in the rectum, then displacing the stone downward, until, once sufficiently descended, it could be extracted by an incision in the perineum. This procedure was certainly not easy to perform and shows his level of surgical sophistication. The rich variety of surgical instruments then in use is well described and beautifully illustrated in his treatise. It contains valuable advice for extracting arrows, setting fractures, and correcting upper limb luxations. Translated into Greek and Latin, his treatise inspired European surgeons of the Renaissance, such as Guy de Chauliac (c. 1302–1368) in fourteenth-century France and Fabricius ab Aquapendente in Italy, three centuries later.

Maimonides (Moshe ben Maimon in Hebrew, 1135–1204) was all at once physician, theologian, philosopher, and spiritual chief of Judaism. Fleeing from anti-Jewish persecution, he traveled between 1148 and 1165 to southern Spain, France, and North Africa, so his contributions to surgery were relatively restricted. Maimonides's chief interest lay in internal medicine. He wrote commentaries on the works of Galen and Hippocrates, enriched by his critical analyses. He devoted particular attention to psychic troubles and the effect of passions and agitations of the soul upon bodily health.

Meanwhile, the situation in Europe could not have been more pitiful. There were no trained surgeons, with the exception of Jews who had studied Arabic medicine. But these men labored under severe restrictions imposed by the Catholic Church. When the famous school of medicine at Salerno[6] was first started, in the tenth century, all the instructors were Jews. An attempt was made to develop surgery, but it was unsuccessful, given their rudimentary understanding of anatomy. The precious Arabic texts had kept alive the flame of knowledge but had also perpetuated the Galenic anatomic misconceptions.

According to the historian Nancy Siraisi, in the Latin West between the sixth and eleventh centuries, little or nothing was added to the knowledge of surgery, which had not yet been differentiated from the rest of medicine.[7] Later, stimulated by the body of knowledge preserved in the Arabic works, Latin-speaking authors began to produce texts of their own. This revival of medical and scientific knowledge in the West must have been fraught with untold difficulties. The Arabic texts had been largely inspired by Greco-Roman works translated into Arabic. Now retranslation caused confusion and misunderstandings, making it difficult to transfer concepts and meanings from one language to another and from one cultural context to another. These were the problems faced by the Salernitan school.

When Vesalius dared to challenge Galen's teachings during the Renaissance, the leading intellectuals berated him. His impugners were ready to defend Galen even in the face of the most glaring contradictions. For instance, when Vesalius demonstrated that the head of the human femur is not flared, stating that the ancient Greek master had used a quadruped's hip for his descriptions, his opponents responded that Galen had not lied and, if the human hip did not conform to his description, it was because men's anatomy had changed. This, they claimed, was due to centuries of wearing tight trousers instead of loose-fitting tunics and togas, as the ancients had done.

The Salernitan school did more than spur the development of European medicine; it shaped an identity for the medical profession. It structured medical learning, which theretofore had been chaotic, irregular, and all too often left in the hands of empirics and quacks. It even produced a code of conduct to which physicians were expected to adhere, thereby conferring some respectability upon the medical profession.

Yet surgeons continued to be portrayed in an unflattering light. The Church, alarmed at the abuses committed by clergymen who practiced surgery, such as charging immoderate fees, and fearful of the harm that patients could suffer from the procedures, issued an edict in the Lateran Council of 1139 condemning priests and monks who engaged in the healing arts for lucre. The restriction went still further in the Edict of Tours (1163), which peremptorily declared that "the Church abhors the shedding of blood" (*Ecclesia abhorret a sanguine*). A priest could not celebrate the Eucharist with bloodstained hands. But by disbarring churchmen—the best-educated class in the Middle Ages—from practicing surgery, this activity was left in the hands of the uncultivated and ignorant. Those were the times of itinerant surgeons: uncouth, loquacious, deceitful, fraudulent men who would perform operations in a town or at a country fair and then quickly abscond for fear of the reprisals that their bunglings were sure to provoke. Little wonder that the Montpellier school eliminated the teaching of surgery, and warned the students to stay away from that suspect field.

In France, three classes of medical men were practicing medicine by the late Middle Ages. The highest class was the physicians: they knew Latin, had benefited from a university education, and were schooled in the subtleties of scholastic philosophy. Their role was to prescribe and to give advice. Next came the long-robed surgeons: they were hands-on practitioners who could dress wounds, set fractures, and apply poultices and plasters but who did not perform incisions or invasive procedures, as they would not be stained with blood. Last were the "barber-surgeons," distinguished by their short robes. Barbers they were, indeed, for their original occupation had been to shave monks, after a 1092 Church decree that forbade monks to sport beards. They were the "untouchable caste," so to speak, of the profession. They could bleed patients, lance boils, and perform other invasive procedures; but they spoke no Latin, had not had a university education, and were regarded by their more exalted colleagues as menials. It took a long time for these distinctions to disappear.

In London, a Barbers' Company was established by ordinance in 1376. A Fellowship of Surgeons was formed in 1365, slowly grew, and received a royal charter in 1462. Gilbert Kymer, the dean of Oxford, petitioned King Henry V to regulate all members of the healing profession and to place them all under the direction of the physicians. Envy and private interests foiled attempts at unification, and disputes between the groups continued for more than a hundred years. In 1540, Henry VIII signed a charter that incorporated barbers and surgeons in a common guild, which would later become the Royal College of Surgeons.

Well into the Renaissance, the original amalgam of surgery and the barber trade still held, so that one of the duties of army surgeons was to shave officers during military campaigns. French surgeons received royal favor, and consequently social prestige, after the successful operation on Louis XIV for an anal fistula in 1687. In England, although the status of surgeons improved, physicians continued to distrust surgeons through to the nineteenth century.

It was the same throughout Europe. With the legitimation of surgery came the need to regulate its practitioners' education. A surgeon's education had been largely an apprenticeship, often attached to a demanding master and under difficult living conditions. Typically, the apprentice was lodged in a dingy garret annexed to the master's house; fed less than liberally; ordered around with little circumspection; reprehended with still less consideration; and awakened early in the morning to shave people, bandage ulcerous limbs, and bleed those who needed it, which, according to current medical opinion, was almost everyone.

Talent and good fortune, however, will allow a man to emerge from nameless obscurity against all logical expectations. Such was the case of the greatest surgeon of the European Renaissance, Ambroise Paré (1510–1590). He was born to a plebeian French family that knew no great wealth. After being apprenticed to several local masters, all of whom, he tells us brashly in his memoir, "were surprised of his rapid progress," he went to Paris to continue his studies as a barber-surgeon. In 1533, he worked at the Hôtel-Dieu hospital, where he learned anatomy and witnessed a plague epidemic, thereby gaining considerable experience. Then, in 1536, he enlisted as an army surgeon. The army did not have a regular medical corps: surgeons enlisted voluntarily in times of war. However, the high officers, usually members of the nobility, could "invite" known, talented practitioners to serve under their banners. Such an invitation could hardly be refused.

In 1536, King François I sent a military expedition to the Duchy of Milan in reprisal for the assassination of one of his agents by Duke Sforza. Paré went with the troops commanded by Colonel-General René de Montejan. This was his introduction to the cruelty of war. At a place known as Pas-de-Suse, during the taking of a citadel, he witnessed a deeply troubling, barbarous act.

He was led to a stable, where three wounded soldiers were lying against the wall, "their faces totally disfigured; they could neither see, nor hear, nor speak, and their clothes still burned from the cannon powder that had hit them." An old soldier asked Paré if there was any way that the men could be cured, and, upon being told they could not, "he suddenly approached them and, swiftly, without anger, he cut their throats. Seeing this terrible cruelty I told him that he was an evil man. He answered me that he prayed God that, if ever he found himself hurt that way, there should be someone to do the same thing to him, so as not to languish in misery."[8]

As an army surgeon in a campaign, Paré made one of his important discoveries. Until then, harquebus-produced wounds were thought to be poisoned. Paré quotes in his works the prevailing medical opinion: "The harm of harquebus wounds derives from the poison that the bullet or the powder carry with them, and much less from the combustion or the cauterization that the said bullet, heated by the fire, makes in the parts that it tears with its violence." The treatment of these wounds was to pour boiling oil (oil of elderberry—a tree or shrub of the genus *Sambucus*—was preferred) through the wound tract and to follow this with cauterization with a red-hot iron.

One day, his supply of oil having been exhausted, it occurred to him to apply to the wounds a "digestive" preparation made of egg yolk, rose water, and turpentine. He could hardly sleep that night, worrying over the possible effects of his experimental treatment. Great and joyous was his surprise when, the next morning, he saw that the patients treated with his preparation seemed stable and were resting comfortably, while the unfortunates who had received the boiling oil and the cautery were febrile and in agonizing pain. Right there, Paré "decided never again to burn the victims of harquebusades so cruelly."

"Gunpowder," someone once said, "invented surgery." After repeated exposure to the ghastly bodily gashes and tears of warfare, Paré made the greatest of his discoveries: a new and more effective way to stanch bleeding. Until then, the standard method had been cauterization with a red-hot iron, "a most cruel and horrible thing to recount, causing extreme pain to the patients." The freshly burned tissues were friable, they broke down, and uncontrollable, massive bleeding often led to fatal exsanguination. And if pain, shock, and blood loss did not kill the unfortunate victims, infection of the devitalized tissues usually would finish them off.

Paré believed that the method he conceived was inspired by God, "without having seen or heard anyone, nor read before, except in Galen's fifth book of his Method, where he writes that one must tie the vessels at their roots, which are the liver and the heart, if one is to check the flow of blood" (he was referring to the Galenic notion that the veins "originate" in the liver and the arteries in the heart). Paré's idea was crucial to the history of surgery. He realized that bleeding sites could be controlled by tying off the severed ends of the blood vessels. From our vantage point, this seems obvious. But we must remember that the circulation of the blood was not yet understood and that knowledge of anatomy was just consolidating. In that context, the ligature of blood vessels was a major advance.

Ambroise Paré was again in a military campaign in 1552. At the siege of the town of Damvillers, near Verdun, the high officers were meeting inside a tent when they became the target of a culverin shot. The sixteenth-century culverin was a long cannon that could fire balls

Ambroise Paré devising the ligature of blood vessels at an amputation.

up to sixteen pounds over a long trajectory. The ball went through the tent and smashed the leg of one of the high commanders. "I had to finish cutting off his leg," wrote Paré, "and this was done without applying the hot irons." By then he had developed a special forceps, graphically named *bec de corbin,* or "crow's beak" (similar to the instruments of the ancient Indian physicians, fashioned in the shape of animals), which made it easier for him to grasp the cut blood vessels that retract after being severed and may hide in the tissues. In this way the surgeon was able to tie them.

While some consider Paré to be the "father of surgery," others bestow that title on the Scottish surgeon John Hunter (1728–1793). Like Paré, Hunter was likelier to favor experience-tested remedies over theory-based precepts and was unimpressed by received opinion. Like Paré, he distrusted bookishness in the medical art, and, lacking a university education, he learned his trade as an apprentice to prominent surgeons such as William Cheselden (1688–1752) and Percivall Pott (1714–1788). Unlike Paré, however, Hunter lived in the Age of Enlightenment, an era when society was more favorably disposed toward learning and the prestige of surgery was on the rise. He contributed in

John Hunter (1728–1793).
COURTESY OF THE NATIONAL LIBRARY OF MEDICINE

a major way to the knowledge of anatomy, dentistry, and clinical medicine, through a number of books that included *Natural History of the Human Teeth* (1771), *The Digestion of the Stomach After Death* (1772), *Treatise on the Venereal Disease* (1786), *A Treatise on Blood Inflammation and Gunshot Wounds* (1794), and others.

John Hunter was a colorful personality with many quirks, including his association with grave robbers, who provided him with cadavers for dissection; his bent toward experimentation, which led him to attempt procedures ranging from artificial insemination to tooth transplantation; and his cantankerous disposition, which set him against many of his contemporaries—all these added a novelesque tone to his life, much exploited by his biographers.[9] He died suddenly, apparently from a heart attack during an acrimonious dispute with his colleagues over who would succeed him in his hospital appointment.

THE DEVELOPMENT OF ANESTHESIA

Vesalius and Paré: each of these two giants razed a major obstacle to the progress of surgery. One charted the interior of the body so that surgeons could keep their bearings at all times. The other, by showing how to avoid catastrophic hemorrhages, emboldened surgeons to try more daring interventions. But this was not enough. Surgeons had not yet found a way to mitigate the pain of a body being sliced with knives, crushed with clamps, rent with scissors, and punctured with needles or to avoid the infection that invariably followed the violation of bodily integrity. The solution to the first problem came with the development of anesthesia.

The struggle against pain is very old. All cultures have used herbal compounds or alcoholic beverages to deaden painful sensations. The Greeks and the Egyptians knew about opium and may have applied it directly to wounds. The Peruvian Incas chewed coca while enduring a trepanation. The surgeon who performed this operation would chew on this plant, too, and some say he would let his coca-saturated saliva fall on the wound. The Chinese had opium and hemp, but, according to Majno,[10] used neither. Instead, they resorted to henbane, a plant of the genus *Hyoscyamus* from which are obtained three powerful drugs: atropine, hyoscyamine, and scopolamine. Scopolamine produces amnesia of the painful experience. It was first synthesized in Germany in the 1890s and then widely used in America, especially in obstetrics, during the early twentieth century. The seeds of this plant (*H. niger*) have been used in India for toothaches; the leaves, in beverages or smoking mixtures.

Efforts to induce anesthesia by inhalation are also quite old and can be traced to the medieval Arabic physicians, who soaked "soporific sponges" in various drugs (opium, mandragora, henbane, alcohol), then had the patient inhale the vapors. This allowed some pain control, but the duration and depth of the anesthesia were probably incompatible with major operations.[11] The thirteenth-century Spanish alchemist and philosopher Ramon Llull (1232/33–1315/16) found that if a mixture of vitriol (sulfuric acid) and alcohol is distilled, it is possible to recover a clear fluid, originally called "sweet vitriol" and later known as ether. Some attribute its discovery to the German botanist Valerius Cordus (1515–1544) in the sixteenth century. Regardless of who originated it, this was a very important discovery. Ether remained in the arsenal of physicians for various treatments, but, amazingly, six centuries passed before it was realized that inhalation of this substance achieves a profound anesthesia.

Meantime, surgery was a ghastly affair. In the early nineteenth century, surgeons donned overalls soiled with gore and filth; their hands were commonly ungloved and unwashed; the patient was conscious, forcefully restrained or tied, and screamed with agonizing pain during the procedure; the operating table was a slab of wood with a channel through which blood and secretions dripped into a sawdust-filled bucket below; and the agonies of the patient (some said, more appositely, "the victim") were watched by dozens of students who crowded a hot, ill-ventilated surgical room.

Little wonder that Charles Darwin (1809–1882), who in his youth wanted to become a surgeon and had actually enrolled in Edinburgh's medical school, was dissuaded from this plan after watching two operations. He left the amphitheater before they were completed, writing, "Nor would I ever attend again.... The two cases fairly haunted me for many a long year."[12] The image of the surgeon was that of a man of resolve and strong stomach, as a sixteenth-century physician-surgeon put it: "nimble handed, sharpe sighted, pregnant witted, clenly appareiled, pitefull harted, but not womenly affectionated: to wepe or trimble when he seeth broken bones."[13] Much of this stereotype has survived to our day in the tribal lore of the medical profession. The surgeon is often depicted as no-nonsense, aggressive, and blustery. William Hunter (1718–1783), brother of the famous surgeon, was of the opinion that surgeons' daily contact with bodies lying at their mercy "may render them less able to bear contradiction."

True, a "womanly affectionate" disposition would have been out of place in a surgeon of those cruel times; and nimble-handedness was a paramount quality, for the atrocious sufferings and the blood loss had to be kept at a minimum, lest the attending shock kill the patient. Hence the need for speed in operating. William Cheselden, a noted English surgeon-anatomist and John Hunter's mentor, was able to perform a lithotomy, the removal of a bladder stone, in only fifty-four seconds, whereas formerly it had taken more than one hour. Baron Dominique-Jean Larrey (1766–1842), a military surgeon with Napoleon's army, was able to complete an amputation in one minute. Improvements in the control of pain, especially the discovery of modern methods of anesthesia by inhalation, would greatly alter these conditions.

The inhalation of gases became somewhat of a fad in the eighteenth century. Joseph Priestley (1733–1804), the English political theorist, clergyman, and scientist and one of the discoverers of oxygen, studied other gases, or "airs," as they were called. Among the "new airs" that he described in *Philosophical Transactions* (1772) was nitrous oxide, which later came to be known as "laughing gas" because of its ability to produce dizziness and giggling. People inhaled it for fun at fairs and parties. They became drunk and often fell but appeared not to feel pain from the trauma of their falls while under the effects of the gas. This led a dentist, Humphry Davy (1778–1829), to try it on himself to relieve the pain of an inflamed gum. He reported that, when mixed with oxygen, nitrous oxide induced temporary unconsciousness and suggested that this property might be useful to dull the pain of surgical procedures. Yet he did not pursue this intriguing idea.

Americans made the next advance, about forty years later. "Exhilarating or laughing gas" was enjoying a vogue in the United States by the mid–nineteenth century. There were "laughing gas" parties and public shows. Horace Wells (1815–1848), a dentist of Hartford, Connecticut, attended one of them and was convinced to utilize nitrous oxide in his practice after trying it on himself. He organized a public demonstration at the Massachusetts General Hospital in which a patient was to have a painless tooth extraction. Under the stern gaze of the eminent Dr. John C. Warren (1778–1856), in whose class the demonstration took place, and a skeptical public composed of physicians and dentists, the patient, a young boy, was set up, the gas was administered, and an extraction was performed. It was a flop. The boy shrieked in pain. This dismal failure proved too much for Wells: his support was withdrawn, his peers made him the butt of ridicule, and he ended a broken man, addicted to narcotics. Jailed for some sordid crime against prostitutes, he committed suicide while in prison.

However, his colleague and former partner William Thomas Green Morton (1819–1868), of Charlton, Massachusetts, was not so easily discouraged. He became a medical student as well as a dentist. Ether, "the most volatile and most inflammable of all known liquids," as a physician called it in the eighteenth century,[14] had many uses in medical practice. Like nitrous oxide, ether became a fad. Medical students had "ether frolics" for amusement, during which they inhaled ether for the intoxication, or "ether jag," it produced. It was a source of inexhaustible fun for people to see other people making fools of themselves. Morton joined in the fun and could not help but notice that, drunk with ether, the students seemed unmindful of the hurts they sustained while thrashing about.

Other American professionals were also aware of the insensitivity to pain that could result from the use of ether. William E. Clarke (1818–1878) had successfully performed a tooth extraction under ether, and Crawford Long (1815–1878), a country doctor born in Danielsville, Georgia, who had himself participated in "ether frolics," administered the substance to a boy prior to excising a cyst from his neck. Doctor Long practiced in a little town, Jefferson, of only a few hundred inhabitants, but he caused quite a stir in his community for his use of ether as an anesthetic, and there is evidence that William Morton visited that town and must have been aware of the much-talked-about experiments. Long was also the first to use ether anesthesia in an obstetrical procedure in 1845.

Anxious to prove that ether was the pain suppressant mankind yearned for, Morton tried it on himself to the point of falling unconscious. Then he organized a public demonstration under conditions designed to preclude all possibility of deception. On October 14, 1846, at ten in the morning, a hard-nosed audience of surgeons gathered in an operating room of the Massachusetts General Hospital, ready to sneer at one more in a series of quacks pretending to revolutionize surgical practice. Morton was late, for he had been directing the construction of an inhaler, keenly sensitive to the fact that the failure of his predecessor, Horace Wells, could have been due to inadequate delivery of the anesthetic. The administration of the anesthetic had yet to be systematized: some advocated as simple a method as soaking the corner of a towel with ether and holding it to the patient's nose. Morton devised a machine with valves, so that the inspired air would carry the anesthetic but the expired air would be detoured, not to dilute or vitiate the incoming ether. Busy with the last details of his machine, he arrived late at the amphitheater, just as the surgeon, the eminent Dr. Warren, was about to make the first incision in the removal of a vascular tumor from the neck of a young man named Gilbert Abbott.

Warren addressed Morton impatiently: "Well, sir, your patient is ready!" Morton applied the anesthesia, the surgeon made his cut, and

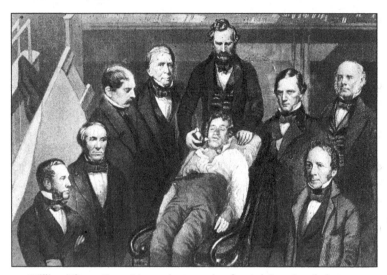

William Thomas Green Morton demonstrating the administration of ether anesthesia at the Massachusetts General Hospital. COURTESY OF THE NATIONAL LIBRARY OF MEDICINE

the patient, sunk in a deep slumber, uttered no complaints at all. The looks of sarcasm, contempt, or derision turned to seriousness and astonishment as the operation proceeded in complete silence. At one point Dr. Warren declared solemnly: "Gentlemen, this is no humbug!" The operation was finished, and, although the patient was somewhat agitated and incoherent toward the end of the procedure, pain had been effectively suppressed. A new era of surgery had dawned.[15]

Soon thereafter, operations were done with complete absence of pain. Several cases were reported in detail in a prestigious medical journal.[16] Worldwide acclaim followed swiftly. Apart from the biting humor of George Bernard Shaw, who declared the advent of anesthesia a disservice to mankind, for it meant that "every fool could be a surgeon," the discovery was saluted with unalloyed enthusiasm. Unfortunately, greed and commercialism came with it. Morton and a close associate, Charles Jackson, descended to imposture and cheap quackery: using dyes to alter the appearance of ether and adding aromatic oils to disguise its characteristic smell, they tried to patent it with the pompous brand name "Letheon." Then they waited to get rich.

Their scheme did not prosper. Dr. Warren and others declared the proposal unethical, since it attempted to deprive mankind of the one benefit—the abolishment of pain—that should be denied to no one. Morton and Jackson had to withdraw their patent application. At first, these two claimed to be the original codiscoverers of ether anesthesia. Their claim was later disputed by William Clarke and reluctantly, prompted by influential friends, by Crawford Long. Eventually, open and rancorous bickering broke out among all the claimants. Incredible as it sounds, the U.S. Congress was dragged into the fight, with senators favoring one or the other of the disputants. Then Morton suddenly died on July 15, 1868, apparently of apoplexy. Congress never ruled on who was the true discoverer, and various associations of American physicians and dentists have since argued that one of their number deserves all the plaudits.

Sir William Osler decided in favor of William Morton, because he was the one to diffuse the knowledge to the entire world. But many point out that Morton could do this only thanks to his position in a prestigious medical center and his access to respected journals of worldwide circulation. Moreover, in 1842, he visited the little town of Jefferson, where Crawford Long was practicing. Were it not for this visit, say Long's advocates, Morton might have forever remained a stranger to the pain-suppressing potency of ether. Jokers suggested erecting a monument with the two, Morton and Long, poised in a sculptural pair and the words "To Either" on the pedestal. The American College of Surgeons endorsed Crawford Long as the discoverer and sponsored the Crawford Long Association, which in 1926 erected a statue to him in Washington, D.C.

Of enormous importance in the history of medicine was the obstetrical application of inhalation anesthesia. Millions of women throughout the world would benefit from it during childbirth. On January 19, 1847, barely three months after Morton's demonstration, a Scottish obstetrician named James Young Simpson (1811–1870) administered ethyl ether to a woman who was having a difficult delivery. Simpson would become a powerful advocate of the use of anesthesia in childbirth. Having found that chloroform, the anesthetic discovered in 1831 by the American chemist and physician Samuel Guthrie (1782–1848), was superior to ether, Simpson gave it to his niece when she was in travail, after testing it on himself. Ether was highly irritating to the airway and the lungs and often provoked persistent vomiting. In addition, accidents and explosions were known to occur in the operating room, because of its highly flammable nature. Chloroform, as it turned out, was also quite toxic: it is now known to be severely injurious to the liver and kidneys, but these effects had not yet been documented. Simpson was supposedly fond of testing anesthetic agents at home, during convivial gatherings with his friends, and during one of these sessions he was accidentally anesthetized by chloroform. Presumably he awoke in a daze, only to find every one of his dinner guests flat on the floor, totally unconscious.

Those who knew him agreed that James Young Simpson's personality was as impressive as his physical presence was unforgettable. He had an entrancing, penetrating gaze; his full mouth was said to be strangely expressive; and his head was qualified by all as unusually large, the effect of volume further amplified by wearing his hair long and often in a tangle. Because of his strong, heavyset frame, it was said that he had "the head of Jove and the body of Bacchus." This was the age of phrenology, and much was made of the large size of Simpson's head: a popular misconception was that cephalic volume correlated directly with intellectual power. When he died, the cranial autopsy findings were reported in detail: skull circumference taken at the occipital protuberance and below the frontal bosses was 22.5 inches; from ear to ear, 13 inches.[17]

Simpson had to contend with a vociferous opposition, led within the medical profession by Charles D. Meigs (1792–1869). Meigs believed that pregnancy and delivery, being natural processes, should not be interfered with. Both sides made strong arguments, but Meigs's staunchly conservative views slipped into arrant reactionism. In his textbook, he stated that a woman's place is in the home, because "her intellectual force is different from that of her lord and master." Harping again on cephalic volume, he said that a woman's head is "too small for intellect, and just big enough for love."[18]

The controversy was raging when Queen Victoria's accoucheur, Sir Charles Locock (1799–1875), decided to try chloroform anesthesia on the royal person in 1853. For this he called in the illustrious John Snow (1813–1858), who would go on to discover the manner of transmission of cholera (1849). Snow had considerable experience with obstetrical anesthesia, which he proceeded to apply by the open drop method (not with an inhaler) to Queen Victoria, who was under this procedure for the fifty-four minutes of the parturition that brought forth her eighth child, Prince Leopold. The queen was delighted: she never lost consciousness and was dimly aware that she was having a child but said she experienced no pain.

The prestigious medical journal *The Lancet* thundered in its issue of May 14, 1853:

> ...great astonishment...has been excited throughout the profession by the rumor that her Majesty during her last labor was placed under the influence of chloroform, an agent which has unquestionably caused instantaneous death in a considerable number of cases. Doubts on this subject cannot exist. In several of the fatal examples persons in their usual health expired while the process of inhalation was proceeding, and the deplorable catastrophes were clearly and indisputably referable to the poisonous action of chloroform, and to that cause alone.

The Lancet's editorialists were right. Chloroform is highly toxic and may cause sudden death. Fortunately, in later years it was superseded by better and safer agents. But the queen underwent the anesthesia without apparent ill effects and repeated the experience four years later with her ninth and last pregnancy, upon the birth of Princess Beatrice (1857). The notoriety of these episodes, and the social status of the patient, were very powerful elements of advocacy for the use of obstetrical anesthesia.

There followed a number of improvements. The inhalers were crude devices in need of standardization. There was a constant danger of administering either too little or too much, even a fatal dose. Ingenious tubing systems were invented to convey anesthetic gas from the machine to the patient while allowing the patient to breathe. Ivan Magill (1888–1986), an Irish-born physician, devised a "breathing system," or "circuit," as it is now called, consisting of a reservoir bag, tubing inserted into the patient's airway, and a pressure relief valve, variants of which are still in use.

The tubes inserted into the trachea, however, did not protect against the danger of aspiration of secretions and sometimes of vomitus. Therefore, the distinguished American anesthesiologist Arthur Guedel (1883–1956) devised cuffed endotracheal tubes that enabled ventilation and suction. He experimented with animal tracheas supplied by butchers and glued inflatable surgical gloves to the tubes' outer walls. Eager to convince his colleagues of the virtues of his inventions, he resorted to an attention-getting maneuver that, although in many ways objectionable, was nonetheless quite persuasive. He anesthetized his own pet dog, quaintly named "Airway," intubated it, and submerged it in a water tank with appropriate ventilatory support in front of a public of anesthesiologists and physicians. Long after the dog would normally have drowned, Airway was removed from the tank and allowed to recover; the dog shook the water from its body and walked away with its usual pertness. The group of professionals were now firm believers in Guedel's procedures.

There came new anesthetic gases. Cyclopropane was widely used in the 1930s and 1940s. Halothane, which was nonflammable, was introduced in the 1950s. Curare, long known for its paralyzing properties, was synthesized in the late 1940s and enlarged the surgeon's armamentarium. When the abdominal muscles were relaxed, the surgical manipulation of visceral organs was much easier: theretofore, merely replacing intestinal loops in an open abdominal cavity had been a nightmare. Curare also paralyzed the chest muscles, thereby making it possible for the anesthesiologist to control the patient's respiration with the appropriate machines, finely adjusting the supplied gases to the exact physiological requirements. Ultra-rapid-action sedative-hypnotics, such as pentobarbital and other barbiturates, were used to induce and maintain anesthesia, and occasionally as primary agents.

To these advances was added the development of local anesthesia. Injection of anesthetic agents into specific nerves made it possible to desensitize portions of the body and thus to perform certain operations without any need to induce unconsciousness. William Stewart Halsted (1852–1922), perhaps the most influential American surgeon ever, was the creator of the operation for breast cancer known as "radical mastectomy" and the first to inject cocaine into nerve trunks, thus pioneering the anesthetic technique of nerve blockage. Unfortunately, the addictive nature of cocaine was not known, and Halsted, who experimented with it, became an addict. Although his addiction did not interfere with his professional life at first, its effects were reflected in his personality: once an extrovert, he turned into a recluse, desperate to find solace in morphine and alcohol.

FURTHER DEVELOPMENT OF SURGERY IN THE POSTANESTHESIA ERA

As soon as surgeons realized that they could control pain, hemorrhage, and infection (see chapter 5) and maintain patients in a good physiologic state with the appropriate supportive measures, their field blossomed. Surgical procedures were attempted that had previously been inconceivable. Unfortunately, there were also abuses: surgical remedies were tried where noninvasive treatments might have sufficed, and surgically correctable "diseases" were fabricated (often, it must be said, with the eager collaboration of the patient). One example is the so-called floating kidney. Many symptoms—flank pain, urinary infections, digestive troubles—were attributed to a kidney that would not stay fixed in its normal anatomical location but wandered about, tending to be displaced downward (nephroptosis). Urologists diagnosed hundreds of patients with this condition. Ingenious surgical procedures were created to anchor the kidney to the surrounding structures (nephropexy). However, this disease, extremely frequent before 1950, was very rarely diagnosed thereafter. Surgeons began expressing skepticism about its very existence ("the most serious complication of nephroptosis is nephropexy," a wag declared), and nephroptosis nearly disappeared from the medical literature. Today, most urologists have never seen a well-documented case of the condition.

(In recent years, there has been a mild resurgence of interest in this disorder. Nephropexy can now be performed by laparoscopy, i.e., via small incisions through which are introduced a miniature video camera and tiny instruments, which the surgeon operates at a distance, while watching the procedure on a monitor. Given the history of this disease, it is unavoidable to think that the reappearance of the diagnosis may be related to the advent of novel ways to do the corrective surgery.[19])

Before the 1970s, millions of tonsillectomies were performed in the United States, but the frequency of this procedure has since been reduced by at least three fourths. Episiotomy, the cutting of the lower vaginal wall and perineum to facilitate the last stage of childbirth, became almost routine in many hospitals and remains so in the early twenty-first century. In America, 32.7 percent of all births involve an episiotomy, in spite of claims that the episiotomy may be worse than the spontaneous lacerations it was supposed to prevent.[20] The rate of elective cesarean birth is also very high: 27.6 percent of all births in the United States, where it increased uninterruptedly between 1989 and 2003. Often it is done for no other reason than the patient's choice, even though some data indicate that maternal morbidity and hospital costs are greater than in normal vaginal delivery.[21]

Surgeons are, in a way, victims of their own success. Public expectations are inordinately high. Surgery has progressed from being strictly *ablative,* that is, limited to removing the diseased part, to *reconstructive* and reparative and has now entered the stage of organ *replacement.* Areas of the body that in the past no one dreamed operable, such as the heart and the brain, are now the subjects of complex surgeries. In the United States, Harvey Cushing (1869–1939), a student of Halsted, operated during his career on some two thousand patients with brain tumors. Before him, the mortality rate approached 50 percent; he brought it down to 8 percent, and thanks to technical improvements it continues to decrease. Vascular lesions of the brain, such as aneurysms and arteriovenous malformations, were long taboo for surgeons; today, they are treated both surgically and nonsurgically with remarkable success.[22]

Advances in understanding of the immune reactions that underlie graft rejection, owed in large part to Peter Medawar (1915–1987), ushered in the era of organ transplantation. The kidney was the first organ to be transplanted: it is relatively easy to access, so tissue samples may be obtained for studies of compatibility and monitoring of the graft; and, although it is a bilateral organ, one is enough to sustain life, thus making it possible to have living donors. Initial efforts fared poorly, because the biology of graft rejection was still unknown. The first operation with long-term success was performed in Boston, when the surgeons Joseph E. Murray (Nobel Prize winner in 1990) and J. Hartwell Harrison used the kidney of a patient's twin as the graft.

Christiaan Barnard (1922–2001), a South African surgeon, transplanted the heart of a young woman into a fifty-three-year-old man on December 3, 1967. (Both were white, allegedly to stave off allusions in the international press to "experimenting on black people," especially in a country that at the time still implemented apartheid.) The American surgeon Thomas Starzl performed the first liver transplant in 1963, but long-term survival was not achieved until 1967. Today, in addition to heart, liver, and kidney, transplantation is carried out with corneas, cartilage, lungs, pancreas, endocrine glands, stomach, spleen, and intestines. Multiple organ transplants are performed, such as liver-pancreas, heart-lungs, and pancreas-liver-intestine; these operations are highly publicized. Thus, *The Washington Post* reported, on September 2, 2005, the case of a Japanese baby who had successfully been the recipient of a six-organ transplant: liver-pancreas-stomach-spleen-small-and-large-intestine.[23] In 2006, a French woman was the first person to receive a face transplant.

FINAL CONSIDERATIONS AND CONCLUSION

Freedom from pain has undoubtedly been one of medicine's greatest gifts to humanity. However, the consequences of this immense boon have been far from straightforward. No one would ever defend physical pain or hope for a return of the suffering that people had to endure in the past. But pain, having been a part of human life since time immemorial, could not be expunged without some disturbing social effect.

Human beings have tried to understand the purpose and meaning of pain; otherwise, suffering would simply be a senseless, absurd thwarting of their lives. Every great human production bears pain's imprint. C. S. Lewis remarked that all religions originated and "were preached, then practiced, in a world without chloroform."[24] For the religious, pain may be a test from on high meant to make the sufferer spiritually stronger. For the nonreligious, pain acquires a meaning if endured for the sake of a cause deemed noble or heroic. In art, the psychological tension of suffering may be related to creativity. Humanists and philosophers have warned that a "paradisiacal" world, in which all our needs were promptly satisfied and all forms of pain canceled, would resemble nothing so much as a dead world.

An exploration of the role of human pain in historical, sociological, and philosophical thought exceeds our purview. The relationship of anesthesia to these large questions was discussed by Donald Caton.[25] Here it is important to distinguish surgical from obstetrical anesthesia.

Surgical anesthesia appeals powerfully to the imagination, since it has made possible prolonged and complex operations such as multiorgan transplants. One untoward consequence of contemporary surgical progress is that human beings are viewed as machines composed of parts that may be replaced and interchanged with one another. Organ transplantation reinforces this notion. It is easy to take the reductionist view that man is, essentially, a machine. The scientific view of a human being as a collection of macromolecules in highly complex physicochemical equilibrium is but the modern equivalent of the eighteenth-century savants' comparison of the body to an horologe with cogs, levers, wheels, springs, and a pendulum. But this conception empties the body—the seat of human identity—of all sense and value; it reduces it to a mere contraption.

If a human being is a machine, a sick patient is a machine out of order. Many health professionals act as if medicine were only a technology that aims to fix the broken mechanism, not the art and science of tending to sick human beings. They are not entirely to blame. The medical school curriculum may include humanities courses that, by revealing the rich complexity of human nature, oppose a one-sided, materialistic outlook. But these courses usually occupy a minimal amount of students' time, and are not very demanding. By far the greatest portion of a student's time in medical school is spent in struggling with arduous courses in which the concept of man as machine appears to be the only valid model.

Anesthesia in obstetrics has also raised important questions. First, some reservations existed about its safety for mother and child. John Snow noted in the nineteenth century that babies delivered under chloroform kicked less strongly, cried more feebly, and grasped the bedclothes with less force than those born without this anesthetic.[26] A polemic followed, but it did not come to a resolution until the mid–twentieth century, when the American physician Virginia Apgar[27] devised a system to assess the respiratory status of a newborn infant immediately after birth and thus to determine which newborns require resuscitator measures. This is the "Apgar score," well known to obstetricians and pediatricians worldwide, which grades such manifestations as skin color, reflex activity, respiratory effort, heart rate, and crying. Thanks to this work, it became abundantly clear that anesthetics given to the mother can cross the placenta and affect the child, whose respiration is depressed, with danger of asphyxia.

Obstetric anesthesia is much safer today than in the past, but by the mid–twentieth century the public was better informed, approaching drugs with more caution. There had been unbridled enthusiasm for anesthesia; now there was a reaction against it. Some women preferred the traditional ways of giving birth, including at home, and realized that much had been lost in the process of "medicalizing" this experience. Home births allowed them to socialize with other women who could help them with the household chores and offer advice. Motherhood consolidated the woman's position at home and strengthened the emotional ties among all members of her family. Moreover, bringing a child into the world was a transcendent experience, for which only poetic metaphors could give an approximation.

None of this was preserved in the impersonal environment of a hospital, where busy nurses and physicians were unable to satisfy the mother's psychological needs. The mother felt "like an object in a factory's assembly line." Drowsy or unconscious, she was unaware that she was giving birth and felt deprived of the meaningful experience of motherhood.

There were also accusations that the medical profession was patronizing, oppressive, and manipulative. Allegedly, physicians placed their own convenience above the needs of their patients; it was the former, not the latter, who determined delivery techniques. In the United States, births by cesarean section (therefore entirely under the control of physicians) went from 4 percent in 1950 to 26 percent in 2002.[28] It used to be axiomatic among physicians that once a cesarean birth had taken place, all subsequent deliveries of that mother should take place by the same route ("once a cesarean, always a cesarean"), lest serious complications supervene. Recently, some have questioned the validity of this assertion.[29]

Physicians answer their detractors by pointing out that the generalization of anesthesia, like the "medicalization" of births, arose mainly out of pressure from the patients. Society clamored for relief of pain and demanded the widest possible application of anesthetics, even when many physicians cautioned against it. Pain is a biological phenomenon. It is a conscious sensation that arises when certain specific nerve impulses arrive at the brain and are "integrated" among various neuronal centers. As a strictly biological occurrence, it can have no connotations of social dominance or oppression. Thus, in medical parlance, "control" of pain and of certain functions (such as respiration under anesthesia) refers to mastery over certain physiological processes. It is essential for a physician to "control" them during surgery or obstetrical procedures. But the term may have been misinterpreted by nonmedical specialists as implying some sort of domineering or manipulative stance on the part of the medical profession.

Today, some women maintain that traditional ways of birth allow them to reclaim the meaning and value that medicalization has taken away, even though they know that there will be pain. Ironically, when anesthesia was a novel procedure and therefore of dubious safety, the public reacted strongly in favor of it, against the caveats of some physicians; now, as anesthesia has become safer, there has arisen a public reaction against it and in favor of traditional ways, against the caveats of some physicians.[30] The obstetrician is therefore compelled to balance each patient's values and preferences against the possible medical risks.

3

VITALISM
AND MECHANISM

For a long time, living and nonliving things were conceived of as separated by an unbridgeable gulf; they belonged to two distinct, irreconcilable domains. Some people argued that nonliving, inert things would never be able to produce the substances elaborated by living beings. Then, in 1828, the German chemist Friedrich Wöhler (1800–1882) succeeded in obtaining urea, the chief product of the metabolism of proteins in the human body and the main nitrogenous compound found in urine, from inorganic reagents in the laboratory. Thus, he showed that an organic substance could be formed from inorganic, inert material (in this case, ammonium cyanate). This blurred the formerly trenchant division between living and nonliving forms and was a defeat for the philosophical school of thought known as "vitalism," as explained in what follows.

At the heart of all inquiries into the origin and maintenance of life, we find two different and mutually opposing viewpoints. One holds that the natural activities of the body are directed by a special force, one that is unique to living beings and that permits them to go on living. "Vital principle" is the most common name for this alleged force, and the philosophy whence it arose is termed "vitalism." Initially, vitalists believed that life is not ruled by the same natural laws that govern nonliving things. They contended that the physical and chemical changes that take place among inert things, as studied in the laboratory, differ fundamentally from what goes on in living tissues. Later, they came to accept that the phenomena of life were exactly the same as those reproduced in the laboratory, but they still maintained that this could not account for the living process. Life requires an immense multitude of physical and chemical phenomena occurring harmoniously, and this effective coordination, according to the vitalists, can happen only if one accepts the directing role of a "vital principle."

Vitalists believed that the vital principle operated in opposition to physical laws. They reasoned that the processes necessary to life could just as easily end it. Living organisms are subject to death and decay. Hence, our bodies would crumble in a matter of seconds were it not for the organizing energy, the vivifying spark, that keeps them going.

Indeed, although the vital principle is difficult to pin down, one definition that was given is "that which opposes the tendency to die."

The contrary viewpoint is called "mechanism," which holds that all forms of life can be explained entirely by physical causes; that life is no more than a special case of physics and chemistry, albeit one of enormous complexity. Unlike vitalists, who believe that biology has its own unique laws, mechanists assert the similarity between phenomena in biology and those in the world of inert matter. In our day, mechanism has triumphed, which has very important implications for medicine.

Some Vitalist Leaders

The origins of vitalism can be traced to Aristotle (384–322 B.C.) and even earlier thinkers. But vitalism was never a well-defined school of thought, and some philosophers may be said to have been vitalists only in the sense of opposing the belief that life could be understood in purely material terms. Aristotle believed that the "soul" animates and directs the body, but his concept of the soul was not quite the same as the vital principle of later centuries. Aristotelian physics proposed a "first motor," itself immobile, which was the ultimate cause of all the movements and changes that occur in Nature. Living beings, in the opinion of the Stagirite, had their own motor principle, the soul, which gave them some autonomy but did not antagonize the universal movement. Thus, Aristotle was really an "animist" more than a vitalist, but this is a subtle and mostly irrelevant distinction.

Vitalism became particularly influential starting in the seventeenth century, and Georg Ernst Stahl (1660–1734), a German physician and chemist, stands as its foremost champion. The son of a Lutheran pastor and attracted to a pietist cult, he was known as an atrabilious, arrogant, morose man. His restrictive, austere religious upbringing created a seemingly dry and melancholy temperament, which in turn made him reserved. Often he did not answer questions (a quality, notes one biographer, that ironically earned him a reputation of great wisdom), did not suffer contrary opinions, and spoke haughtily when condescending to respond. However, another biographer says that these claims were greatly exaggerated. It is known that he married four times, two of his wives died from parturition, and he was survived by several children.[1]

What is certain is that he was a poor writer. His prose was convoluted, turgid, often in Latin sprinkled with German terms, and weighed down with alchemical symbols and esoteric terms with Latinized endings. Obscurity may be the reason why his contributions are known today by so few. There are very few English translations of his works, and those in other languages are just as scarce and difficult to access.[2,3] Yet his influence was widespread and very strong in his time. He became physician to Johann Ernest III, duke of Saxe-Weimar, and professor of medicine and chemistry at the

Georg Ernst Stahl (1660–1734).
COURTESY OF THE NATIONAL LIBRARY
OF MEDICINE

recently constructed University of Halle in 1694. His renown earned him an appointment as physician to the king of Prussia in Berlin, a post he kept until his death in 1734.

Stahl is one of those scholars who become famous by virtue of their indelible association with a wrong theory. He introduced the concept of "phlogiston," which in his writings most often appears in Greek, φλογιστόν, and may be translated as "flammable." It was supposed to be the principle of combustibility: "the matter and principle of fire, not fire itself." Phlogiston, then, was contained in all combustible bodies and in metals, but it had no color, no odor, no taste, and no weight. By means of this strange theory, Stahl set out to explain a number of disparate observations about the combustion, respiration, fermentation, putrefaction, and phenomena observed in metallurgy. When a body burns, he said, something is lost: phlogiston escapes into the air. However, air can receive phlogiston only to a limited extent. Once the air's capacity to admit phlogiston is exhausted, combustion ceases.

In reality, Stahl had inverted the terms: when combustion occurs, the element of combustibility, oxygen, is removed from the air, not added to it, and goes to the burned object. It was already known in his time that some substances, such as magnesium, weigh more after they are burned than before. Stahl, however, disregarded this and other pertinent observations. Followers of his ideas, in their zeal, went so far as to propose the bizarre hypothesis that phlogiston had a negative weight. It took the illustrious French chemist Antoine-Laurent Lavoisier (1743–1794) to topple definitively the fanciful concept of phlogiston when he demonstrated, in his Memoir "On Combustion in General" (read on September 5, 1777, but not published until 1780), that a body can burn only in the presence of oxygen.

Lavoisier's work was inspired by his correspondence with Joseph Priestley (1733–1804), the English chemist and theologian who had discovered the respiration of plants and experimented with that of animals. Another noteworthy predecessor in these studies was the English physician and physiologist John Mayow (1640–1679), who, about a century before Priestly and Lavoisier, had succeeded in identifying a component of air, *spiritus nitro-aereus,* corresponding to oxygen. This man's remarkable insights into respiration correctly identified this process as an exchange between air and blood.

In his Tractatus quinque medico-physici, *John Mayow (1640–1679) wrote of "nitro-aerial spirit," later identified as oxygen, which he discovered when he was only thirty years old. The book was published in 1674.*

Despite his role in the construction of the phlogiston fallacy, Georg Ernst Stahl called people's attention to the chemistry of the human body in health and disease. He believed that the organs of the body were subject to the laws of physics but proposed that the soul regulated and harmonized their functions. Thus his vitalism, like Aristotle's, was a form of animism.

The implications for medicine were very important. Since the soul governed the physiological phenomena, the conditions of the soul were the cause of health and illness. Thus, much attention was devoted to the "passions" that agitated the patient's soul—an attitude at odds with the prevailing one. German physicians of Stahl's time were known sometimes for the insensitive *Rosskuren* (drastic treatments) they were wont to apply. Convinced that medicines exert their proper effect only when the soul is in a tranquil state, Stahl favored psychotherapy preceding or accompanying the administration of medicaments.

Like all physicians of his time, he did prescribe bloodletting (he even recommended doing it twice a year in healthy persons, as a prophylactic measure), laxatives, enemas, diaphoretic preparations, and sundry other therapies that today seem questionable. Nevertheless, his belief in the presence of a regulatory and harmonizing "soul" that supervised the bodily functions had a beneficial consequence. Naturally, he was led to conclude that this "soul" also had the power to heal. From this, he revived the ancient thesis of *natura medicatrix* (Nature as physician), which he reinvigorated for the eighteenth century. Since there is a healing principle inside the body, the medical man should simply be an assistant to Nature's curative potency, not a mechanic repairing a broken machine. Stahl's advocacy of prudent expectation surely had a greater benefit to public health than the combined, misguided efforts of many of his colleagues.

Another important concept in Stahl's thought was what he called "tonic motion" (*motus tonicus*). This was a movement of alternative contraction and relaxation that he attributed to all tissues, except bone.[4] By this motion, blood was propelled to the sites where it was needed. At that time, the circulation of the blood, discovered by William Harvey in 1628, was already generally accepted; but uncertainty remained, and therefore Stahl, like other contemporaries, tried to elucidate it. The presumed tissue contractions expelled the blood from certain territories, while dilations rendered other tissues able to receive the flow of blood and humors. In today's terminology, we might say that Stahl was speaking of "regional circulatory regulation," as opposed to the general circulatory circuit, which had been taken for granted since Harvey's discovery. The soul knew where blood and humors were needed and directed the flow accordingly.

As a chemist, Stahl knew that processes of fermentation and putrefaction can take place inside the body. This was the subject of his first book, *Zymotechnia fundamentalis seu fermentationis theoria generalis.* Why does the living body not decompose or break down? His answer was, again, that the soul prevents it from doing so. Through the agency of "tonic motion" the flow of blood and bodily humors is ensured; and the soul directs this flow to the excretory organs (kidneys, gastrointestinal tract, and others, which he conceived of as "porous"), where toxic substances are removed. Thus, Stahl's vitalism allowed for the obedience of bodily functions to physical and chemical laws, while still proposing that a soul was necessary at every moment if life were to continue.

Stahl's ideas had great resonance throughout Europe, but perhaps nowhere greater than in France, where several physicians from the illustrious school of medicine at Montpellier evolved their own brand of vitalism. Among these, François Boissier de la Croix de Sauvages (1706–1767) was the first to adopt the vitalist notions, to the great agitation of the Montpellier learned community, which until then had professed a thoroughgoing mechanist creed.[5] Sauvages's ideas took especially strong hold on two of his disciples, Théophile de Bordeu (1722–1776)[6] and Paul-Joseph Barthez (1734–1806),[7] before reaching their highest expression in the figure of Marie-François-Xavier Bichat (1771–1802).

Bordeu proposed that there was a "sensibility" that belonged to the material from which all living structures are made (as a vitalist concept, this "sensibility" was not reducible to physicochemical terms). This idea clashed with the mechanist theory that saw the body as a conglomerate of inert parts cleverly assembled, like a clock in which the gearwheels keep turning as long as it is wound. Of course, Bordeu's idea was not entirely original. The Belgian alchemist, physiologist, and physician Jan Baptista van Helmont (1579–1644) wrote of a principle intrinsic to the bodily structures that, in his view, guided and constructed the body. Unfortunately, van Helmont's writings are a tangle of astrological imagery and obscure alchemical references, making them nearly impenetrable.[8] From what sense can be made of his writings, modern interpreters conclude that van Helmont thought this organizing force, which he called *archeus*, belonged to the body itself, controlling all the functions of life via a number of secondary *archei* present in the organism.

Bordeu must have been impressed by these ideas, for in his own discourse on the "sensibility" of each organ, he allowed for the existence of a "general" sensibility, which allegedly exerted overall control. Essentially, Bordeu proposed that each organ had its own individual life, distinguishable from the life of the total organism. His metaphor for this phenomenon was that of a swarm of bees all massed together, hanging from a tree branch: one member of the swarm is attached to its neighbor and the next one to the next, until the whole colony seems to form a solid body. Each bee leads an individual life, but that life is dependent on the lives of the others, and so they all act in concert to maintain the integrity of the entire colony.

Another of Bordeu's precursors was the English physician Francis Glisson (1597–1677), Regius Professor at Cambridge, whose name every student of anatomy learns to associate with the capsule of the liver, to this day known as "Glisson's capsule." He formulated the concept of "irritability" in his treatise on the stomach and intestines (1677) and saw this not only as the primary cause of muscle contractility but as a general property of all human tissues, which he believed was composed of innumerable delicate fibers.[9]

It is not difficult to see how these ideas on "irritability," "sensibility," and the like may have originated. Anyone familiar with a biology laboratory or the receiving room of a pathology laboratory or any place where freshly removed organs and tissues are examined knows that movement and contractility are the most striking hallmarks of life. It is quite impressive to see a heart, long after its removal from the thorax of an amphibian, continue beating with its own rhythm; or a segment of intestine, after it is removed by the surgeon from the abdomen of a human patient, evincing wavelike contractions in the stainless steel pan in which it is placed for examination by the pathologist. Such startling sights must have inspired vitalists with the idea that life resides in the constituents of the body and is an integral part of their being.

Glisson was also a forerunner of Albrecht von Haller (1708–1777), the Swiss polymath—botanist, poet, politician, anatomist, physician, editor, physiologist—whose encyclopedic knowledge and volcanic energy excited the admiration of his contemporaries. Haller wrote more than 14,000 letters, collected an immense repository of bibliographic references, made meticulous anatomical dissections (a circle of minute blood vessels in the ocular globe bears his name, "Haller's ring"), described the flora of the Swiss Alps, wrote passable poetry, disputed with leading intellectuals of his time, and still managed to get married three times and have eleven children. But the most important part of his work was in physiology, which he saw as *anatomia animata,* the study of the movements that agitate the various parts of the body.

Like Glisson, Stahl, and later Bichat, he thought of the fine constitution of the body as made up of fibers. And the movements they manifested he interpreted as reactions to stimuli. He distinguished between sensibility and irritability: a body part was irritable if it contracted following some irritation; it was sensible if it transmitted to the soul the impressions it received. After numerous experiments with all sorts of stimuli, such as heat, electricity, chemical compounds, and so on, he concluded that sensibility was a property of nerve tissue only and irritability, of muscle alone.[10] Von Haller was apparently subject to depressive bouts. Historians speculate that, being of a deeply religious temperament, he may have experienced some remorse for the suffering he inflicted on animals in his experiments on sensibility and irritability. This supposition is based on the numerous apologies in his treatises, where he justifies the apparent cruelty on the ground that humanity would benefit from his discoveries.[11]

The significant point is that these various systems all proposed that the life energy, which most researchers felt incapable of describing other than in metaphors—flame, spark, power, fire, *archeus,* vital principle—was an integral

Albrecht von Haller (1708–1777).
COURTESY OF THE NATIONAL LIBRARY OF MEDICINE

part of the tissues and organs. An immortal soul could exist (in fact, everyone took this to be undeniable truth), and this soul conferred reasoning, intellect, memory, and imagination, but not life. Life was constituted of the bodily structures.

In contrast, the mechanists, influenced by the philosopher René Descartes (1596–1650), believed the body to be a contraption made of nonliving parts and the soul something superimposed on it—such was the central tenet of the so-called (and much maligned) Cartesian dualism. The soul was one and indivisible; it had to do with the noblest functions of human beings, namely reasoning and intellect. Note, however, that the "passions," such as anger and lust, depended on the body, not the soul: they were utterly corporeal. Their promptings sometimes placed human beings in conflict with themselves, when soul and body showed contrary tendencies. Yet the soul came from outside, "infused" on the body as an external agency; and without it the body remained a pure mechanism of lifeless gears, levers, and rotating cogs and wheels, much like a robot or a mechanical toy. Such was the case with animals, which, having no souls, were pure automata.

Barthez, the other great vitalist of Montpellier, appeared after Bordeu but surpassed him in influence and renown. He gave a discourse on the "vital principle" in 1772. However, his vitalist system lacked coherence and contained some blatant contradictions. On the one hand, he assumed the "vital principle" to be unitary, severely criticizing those who, like van Helmont and Bordeu, had favored a plurality of vital powers. On the other hand, he seemed at pains to explain why some parts of the body continue to move even after they have been extirpated from the organism. He proposed that they retained a part of the vital principle, but this contradicted the tenet of its indivisibility; or else the vital principle was extinguished gradually, which was also contrary to his earlier pronouncements about its inalterable oneness. By the same token, he could not define clearly whether the vital principle was separate from the material body or joined to it permanently or impermanently.

While Stahl's writing style had been baroque and torturously rhetorical, van Helmont's cryptic and undecipherable, and Bordeu's equivocal and perplexing, Xavier Bichat wrote clearly and precisely. He expressed his ideas in a simple and comprehensible manner, which is why he became known as the leading advocate of the vitalist persuasion; he is now known as the vitalist par excellence.

Bichat was described by his contemporaries as an amiable man of middling stature and piercing eyes. He died young and never married. His capacity for work was astounding. He lectured, performed dissections, experimented with animals, practiced medicine, and wrote extensively. His talent for clarity of expression no doubt accounted for the wide diffusion of his works, of which he produced many. His definition of life was celebrated by the leading philosophers and intellectuals of his time. He wrote, "A definition of life has been sought in abstract considerations; it is to be found, I believe, in this brief outline: 'Life is the set of functions that resist death.' "[12]

Today Bichat's system is obsolete. He conceived of two lives in human physiology, which he termed "animal" and "organic." Sensibility and voluntary motor activity belonged to "animal life," since animals evince these functions. The human thinking soul related only to animal life (*anima* being Latin for the soul). Vegetative life (nutrition, respiration, excretion) was "organic" life, because its functions, properly arranged by the vital powers, constituted the organization that maintained life. "Organic," in Bichat's system, alludes to "organization," and this organization resulted from the "vital powers" that his vitalism posited. In the course of his exposition, Bichat contradicts himself and runs into impasses when trying to reconcile his proposed bipartition. Life was not just the outcome of an appropriate organization, as in the case of a machine, but the organization itself was caused and sustained by the vital powers.

Bichat proposed that the phenomena of life were excessively variable and impossible to analyze mathematically, because they were constantly changing. In this they differed fundamentally from purely physical phenomena. One could calculate the speed of a projectile, trace the orbit of a comet, or measure the viscosity of a fluid; but to know with accuracy the amount of blood that goes through the lungs, the power of a muscle, or the speed at which the blood flows was something else. These vital actions, he declared, could not be subjected to the kind of calculations from which a general formula may be derived. Life's ever-changing quality from moment to moment, and from place to place in the body, would forever put vital phenomena beyond the reach of total scientific comprehension.

Bichat is also credited as the founder of histology, the science that studies the organization and structure of the various tissues that make up the bodily organs. For a man who hardly ever used a microscope, this is clear proof of his uncommon genius, but this aspect of his accomplishments will be referred to later in this book.

He may have driven himself too hard. He suffered an episode of hemoptysis while lecturing. Many hours of work in insalubrious conditions predisposed him to tuberculosis. Coming down a staircase at the Hôtel-Dieu, where he had been busy for long hours in the autopsy room, he sustained a fall. Bichat did not recover from this accident: an infection led to a fever, then to a coma, and he was dead fourteen days after the onset of the illness. One of his eulogists said that Bichat "pur-

Marie-François-Xavier Bichat (1771–1802).
COURTESY OF THE NATIONAL LIBRARY
OF MEDICINE

chased learning at the expense of the richest soil of human happiness...his greatest fault was an activity of mind disproportionate to his strength." He died on his thirty-first birthday.

SOME MECHANISTS AND THEIR CAUSE

Descartes is central to the history of the mechanist interpretation of biology. Although he became known primarily as a philosopher, he was much given to physiological studies and even practiced vivisection, as seen in his correspondence from Holland. Tradition has it that a gentleman came to visit him one day and asked to see his library. In response, Descartes led him to a place where he kept a calf, pointed to the animal, and said, "There's my library; for that is the study to which I apply myself now."

In his *Treatise on Man* and his *Principles of Philosophy*, he developed his ideas about the structure of the human body. He compared the nerves to the pipes of a cleverly arranged fountain and the muscles and tendons to the levers and springs that move them. The "animal spirits" (*esprits animaux,* a subtle fluid circulating in the nerves and acting as intermediary between the senses and the brain—it may correspond to today's "nervous impulses") he likened to the water that activates a mill or some other device of continuous movement. The ingenious mechanical arrangement in man's body was presumably devised by God. But, he added, "We see clocks, artificial fountains, mills, and other similar machines which, even though they are only made by men, have the power to move of their own accord."[13] How much more complex the disposition of the parts, and how greatly elaborate the actions that one can expect of the human body, whose design issued directly from the Almighty.

Unfortunately, Descartes's physiology was riddled with mistakes that may be attributable to his invention of facts. One of the world's most eminent philosophers seemed to have forgotten, say his critics, the very rules he created. His penetrating intelligence advocated systematic doubt and the suspension of judgment when facts are not duly corroborated by sufficient evidence. But these rules, so valuable in his metaphysical disquisitions, were neglected when he was in pursuit of subtle questions of human physiology.

The mechanist philosophy gained popularity with some of Descartes's contemporaries. The concept that the body had to be studied as one studies a mechanism had been growing stronger since the Renaissance. Science had to be framed in ever more exact terms, and this required a verifiable quantitation of size, shape, weight, and motion, the so-called primary qualities of objects, as opposed to those qualities that we perceive with our senses, such as color, scent, and texture, which Descartes had denounced as eminently untrustworthy.

There had been physicians who believed that progress in medicine depended on the introduction of the exact sciences to their field of study. One was the Italian Sanctorius (1561–1636), of Padua. Although in his medical practice he clung to the Galenic system and the theory of humors, in his research he was remarkably progressive, always favoring observation and experiment over authority. He designed a measuring apparatus, which consisted of a large, movable platform attached to a steelyard scale, and he used this to quantify changes in the body weight of patients. Meticulously recording the weight changes that occurred after food ingestion, drinking, and excretion, as well as alterations in ambient temperature, he demonstrated conclusively that the total weight of the body is inferior to the expected weight after the ingesta and that a significant loss of weight normally takes place through "insensible perspiration." Thus he is rightly considered the "father of basal metabolism." In his efforts to bring the scientific method to the patient's bedside, he also invented instruments to quantify the pulse and to measure bodily temperature, as well as a hygrometer. Many of his observations were published in his book *De statica medicina* (1614).

If the body was a machine, as Descartes maintained, what sort of a machine was it? Some scholars, profoundly impressed by the Cartesian metaphors, imagined it to be like a hydraulic engine. The movement of fluids, they thought, best characterized the process of life. Life was at its core a highly complex series of liquid currents; and the fluids themselves varied in complexity from coarse and tangible, like the blood, to airy and unperceived, like the "animal spirits" that carried their messages throughout the organism at amazing speeds. Still, the body was a mechanical contrivance, and its proper study necessitated adhering to the physical principles of statics and dynamics. Physiology was largely reducible to mechanics. This was the base of the school of iatrophysics.

Among iatrophysics' main representatives was Giovanni Alfonso Borelli (1608–1679), an Italian astronomer and physiologist. In addition to studying the orbits of the planets, he endeavored to explain the muscular movements and other bodily motions of man and animals (*verbi gratia:* the movement of wings in birds and fins in fish) by reference to mechanical principles. His best-known work, *De motu animalium* ("On the Movement of Animals"), deals with this topic. His analysis of the lever-and-force components of voluntary muscle action, in which his astronomer's knowledge of physics and mathematics came in handy, remains exemplary. Much less effective were his efforts at interpreting cardiac activity and intestinal peristalsis.

Opposite the school of iatrophysics stood that of iatrochemistry, whose votaries were convinced that life was best defined in chemical terms. Jan Baptista van Helmont, although his philosophy was vitalist and his cast of mind not far removed from that of an alchemist (the Inquisition persecuted him under the accusation that he practiced "magic"), was nevertheless interested in the quantifying of physiologic phenomena. He compared the respective weights of water and urine and discovered the specific gravity of urine, a clinical test still relevant in our day. He introduced various medicaments, including some containing mercury, qualifying him as a forerunner of iatrochemistry.

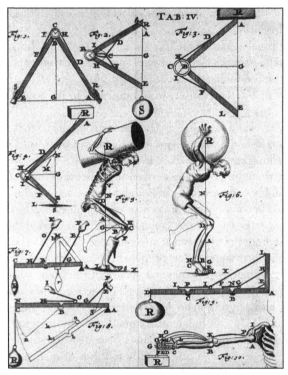

A page from a treatise by Giovanni Alfonso Borelli (1608–1679) on the movement of animals, showing his attempt to analyze the movements of the muscles according to the laws of the science of mechanics as understood in his time. COURTESY OF THE NATIONAL LIBRARY OF MEDICINE

Some say the "father" of the iatrochemical school was the German-born physician Franciscus Sylvius (1614–1672), one of the outstanding European teachers of medicine in the seventeenth century. He is credited with the construction of the first university chemistry laboratory at Leiden, where he taught medicine from 1658 until his death. His name is associated with the deep cleft that separates the temporal lobe of the brain from the frontal and parietal lobes above it (the "Sylvian fissure"). He believed that normal life and disease are explainable as chemical phenomena. He recognized the presence of acids and bases in the blood and thought that their excess—acidosis and alkalosis, respectively—was the cause of diseases, for which he devised appropriate medicaments.

Yet even though evidence steadily accumulated from the early seventeenth century that biological processes were entirely explainable in terms of physics and chemistry, vitalism continued to dominate medical thinking well into the nineteenth century. A serious criticism was formulated early in that century by Lamarck (full name Jean-Baptiste-Pierre-Antoine de Monet, chevalier de Lamarck, 1744–1829) in his main opus, *Philosophie zoologique* (1809). He cogently developed the idea that the laws of chemistry and physics have universal validity: they apply to the whole of Nature, to living beings as well as to inert objects. He argued that there is no such thing as a vital principle preserving living beings from decomposition. The chemical changes of fermentation occur equally in the living and in the dead, but in living beings a normal, healthy organization is present, by virtue of which the chemical reactions yield complex substances that contribute to the maintenance of life. There was no need to invoke a vital principle. In disease, the organization is disturbed; and in death it is completely lost, so that the chemical reactions are misdirected in such a way that decomposition does take place.

As to the question of how a sound organization appeared in the first place, most savants believed that it was there from the beginning. In other words, the organization of living beings was put into place spontaneously, since at the time the majority still endorsed the notion of spontaneous generation. Lamarck believed that the first living beings had been progressively modified by external circumstances and that these *acquired* changes were inheritable, thus accounting for the evolution of the species. Darwinism would discredit this theory a few years later (except in the Soviet Union, where T. D. Lysenko,[14] for strictly ideological reasons, perpetuated an aberrant scientific trend into the twentieth century, thereby setting back Russia's science of genetics for about half a century).

Lamarck's genius as a biologist is undeniable; his studies on invertebrates framed the study of those creatures for decades after him. But he seemed hounded by an obscure destiny. As happened to others, his name became linked to an erroneous theory—"Lamarckism," the inheritability of acquired traits. This sad fate seemed to mirror his private life: he died old, blind, poor, a widower, having seen his major theory contradicted, and forced to go, well into his years, to the French Institute (the foremost French scientific institution at the time) to draw the vouchers that allowed him to eat.[15] No one paid attention to his work anymore, and therefore his criticism of vitalism made no impression.

The major blow to vitalism was dealt by two renowned figures: first François Magendie (1783–1855); then his illustrious student Claude Bernard (1813–1878). Magendie's chief claim to immortality is his ascertainment of the functional nature of the spinal nerves. He confirmed and greatly expanded the observations made eleven years before by the Scottish neuroanatomist Charles Bell (1774–1842) and definitively established that the anterior roots of the spinal nerves carry motor impulses to the muscles, whereas the posterior roots convey sensation.

Magendie was a firm advocate of the experimental method. He reproached the vitalists for the excessively speculative nature of their theories. As his student Claude Bernard put it, "The character of M. Magendie may be summarized by saying that he was horrified by reasonings and theories; he just wanted to see; he always wanted only the facts; he wanted to see only; which is something he himself expressed by saying that he had eyes but no ears."[16]

Whereas Bichat had propounded the great variability of the vital properties, Magendie retorted by saying that such variability, if it were real, would in effect render scientific experimentation impossible. In fact, Claude Bernard later organized his relentless demolition of biological vitalism by pointing to the *constancy* of the inner milieu. Life, he clearly showed, is possible only when the physicochemical characteristics of the internal milieu are rigorously maintained within very narrow limits.

Thus we arrive at our own epoch, which has seen what has been touted by many as the complete triumph of the mechanist conception of life. What seemed at first an insoluble enigma—explaining life and evolution in physicochemical terms—became understandable when the team headed by Oswald T. Avery (1877–1955) reported in 1944 the first experimental evidence of the genetic material of living cells, DNA. The mystery of heredity began to unravel, and, like evolution, it could be wholly explained by the interaction of the molecules that take part in it.

Mechanism, therefore, progressed from iatrophysics to iatrochemistry, to what we might call "iatromolecular biology." As Jacques

Monod (1910–1976) declared in his Nobel lecture, the goal of molecular biology is "to interpret the essential properties of organisms in terms of molecular structures." This goal has been reached in the area of heredity since the discovery of DNA and RNA. But Monod reaffirmed his "confidence in the development of [molecular biology] which, transcending its original domain, the chemistry of heredity, today is oriented toward the analysis of the more complex biological phenomena: the development of organisms and the operation of their networks of functional correlations."[17] The "vital principle" had at last been expelled.

CONCLUSION

Herbert Spencer (1820–1903) wrote, "While many admit the abstract probability that a falsity has usually a nucleus of verity, few bear this abstract probability in mind when passing judgment on the opinion of others."[18] This statement is especially true of the modern attitude toward vitalism. There is no question that vitalism (understood as a preternatural "something" that opposes physicochemical forces and is able to direct them to some end) was superseded, and ultimately reduced to irrelevancy, by progress in the life sciences. But its critics have gone too far in reviling what was once a worthy contrary opinion. Reviewers of historical theories in medicine either ignore or caricature the sensibility that gave rise to the vitalist philosophy. In the old debate between vitalism and mechanism they see a clash between ignorant obscurantism and resplendent truth. One is equated with the animism of shamans or the primitive beliefs of savages; the other is presented as reason triumphant over the retrograde forces of superstition.

The truth is that these two positions were not as sharply divergent as is commonly supposed. A reading of primary sources shows that vitalist authors sometimes adopted postures congenial to mechanism; and conversely, mechanist advocates were rarely so thoroughgoing that their system would not admit some elements of vitalism, especially during historical epochs when there was still much mystery surrounding physiological functions. Of course, today vitalism has no place in science. Perhaps the strongest criticism that has been leveled against it is that vitalist principles were never needed to solve problems related to physiology.

Not only is the mechanist outlook useful in exploring and interpreting the phenomena of life, it is indispensable for this purpose. Yet problems arise when this philosophy attempts to surpass the bounds of biological science and medicine, to become a true metaphysics, an explanation of the whole universe. In that regard, mechanism is on a par with vitalism: they would both be unprovable.

A complete discussion of the current philosophical debate is beyond our purview. Suffice it to mention Jacques Monod's book-length essay entitled *Chance and Necessity*, a forceful manifesto of contemporary mechanism.[19] The book caused a great sensation in intellectual circles when it first appeared and was translated into many languages. Monod, a 1965 co-winner of the Nobel Prize in Physiology or Medicine (shared with François Jacob and André Lwoff) for his studies in bacterial genetics, was also an eloquent writer. In his essay, he proposed that the origin of life and evolution were the result of chance. In other words, life is the result of the gratuitous play of physicochemical forces and does not result from a conscious plan (or, as is fashionable to say now, "intelligent design"). However, since there are no final causes in Nature and no purposefulness to our makeup, and since all matter is similarly indifferent to our self's existence and destiny, this philosophy seems like unmitigated pessimism.

This was fully recognized by Monod, who wrote that the need to abandon the "delusion" that a soul animates our being, and that our makeup reflects a conscious project, is the source of deep anguish. For the lack of such beliefs imposes a radical revision of our ethics; we would have to learn to live morally and affectively without them. Yet the only kind of explanation capable of soothing people's anguish is "one that reveals the meaning of mankind while granting to it a necessary place in Nature." But in order to seem "true, significant and soothing," an explanation of life and our place in the world must perforce join the animist tradition. Throughout history, all religions, almost all philosophies, and "even a part of the sciences" have desperately tried to deny our own contingency. Mankind is seen as part of a majestically unfolding program, like a warp running through the weft of the universe.

Monod concluded that, although objective analysis forces us to see the apparent dualism of our being as false, it is so intimately attached to being itself that it would be futile to hope ever to dispel it. Then, in a surprising turn of thought, he wrote, "To renounce the delusion that sees in the soul an immaterial 'substance' is not to deny its existence, but, on the contrary, to begin to recognize the complexity, the richness, the unfathomable depth of the genetic and cultural heritage, and the personal experience, conscious or otherwise, all of which together constitute the being that we are, sole and unexceptionable witness of itself."

As I understand it, this means there is so much richness and complexity in our beings that we should not lament the loss of the soul. But this is, in a certain sense, what vitalism in its modified form argues. Claude Bernard stated that, in the living being, elements and phenomena do not simply associate with one another but the unification "expresses something more than the simple addition of their separate properties." He went so far as to say that he would agree with vitalists if they would stop invoking the supernatural and confine themselves to "admitting that living beings present manifestations absent in the inanimate world, and for this reason constitute a peculiar character to it."

Two present-day medical essayists, after quoting these statements, conclude, "If it is true that a soul that goes in and out of the organism ... and that opposes and suspends the physico-chemical laws does not really exist, it is equally true that in the whole we call organism, where an enormous number of cells exchange their messages and react in a coordinated way with the purpose of preserving themselves, there exists a common principle, i.e., a unitary and finalistic organization of the vital functions."[20] This principle, they point out, is exactly what Aristotle called the soul. Seen in this light, the quarrel of vitalists and mechanists does not seem so irreconcilable.

The complexity of our being offers new frontiers in biology, according to Monod. Not evolution and not development: we already have the key to these mysteries. Certainly, there remains much to investigate in these areas, but the enigmas they once posed have been solved in principle. Now such questions as the origin of life in the biosphere and the nature of consciousness in the central nervous system are the new riddles that challenge biological scientists.

How many of these riddles will be solved remains to be seen. But we can be sure that the more we find out, the more we shall come face to face with further enigmas. As the known widens its radius, the unknown multiplies its challenges. The brighter the light is made, the greater the density of the surrounding darkness. Vitalists may have been thinkers who concentrated all their attention on the fringe between light and darkness. Their thoughts converged on the transcendent and the insoluble. Surely, they were not all ignorant witch doctors or quacks or shamans: some of the brightest minds were found in their ranks.

A contemporary philosopher, Michel Onfray, questions the scorn that has been heaped upon the "vital principle."[21] This concept, which today has been dismissed, he finds related to an enduring idea, manifested under varied forms, such as the *logos* of the pre-Socratics, Spinoza's *conatus*, Schopenhauer's will to live, Nietzsche's will to power, and other equally respectable members of a very old philosophical tradition.

Vitalism, although useless for a scientific, factual, experience-based understanding of the body, may have been helpful in stimulating intuitions or presentiments, which sometimes turn out to be unexpectedly fecund in science. Philosophically, a vitalist could still argue that we have no concept of life itself, for our much-vaunted knowledge refers only to its outward manifestations. To quote again Spencer (whom Monod placed squarely in the vitalist party): "The explanation of that which is explicable does but bring into greater clearness the inexplicableness of that which remains behind.... [The scientist] learns at once the greatness and the littleness of the human intellect—its power in dealing with all that comes within the range of experience, its impotence in dealing with all that transcends experience."[22]

4

THE MYSTERY
OF PROCREATION

Human generation was the last redoubt of vitalism. Long after the mechanist party had convincingly demonstrated that all the essential phenomena of a living organism were reducible to physicochemistry, the awe-inspiring reproductive function remained unaccounted for. If any bodily property seemed to require the intervention of the "vital principle," surely this was it. For how, if not by invoking a preternatural influence, could one explain the astounding emergence of a human being from the admixture of male and female secretions? Until the seventeenth century, no significant progress had been made in the understanding of this process. It may be said that the prevailing ideas were much the same as those set by the sages of antiquity, especially Aristotle.

The female, Aristotle maintained, contributed only the passive material of which the embryo was fashioned, and this, he thought, was chiefly in the form of menstrual blood. The father provided "form." (In Aristotelian terminology, this does not mean simply "shape" but also the principle that guides the formation of the new being.) Everyone could see that a distinct material, semen, was passed on by the father during the generational act. But this was irrelevant. This fluid, as anyone could see, was not retained by the woman. What really mattered was the embryo-shaping principle, the creative faculty, i.e., Aristotelian "form," carried therein.

In the Middle Ages no further light was shed on the enigma. Progress consisted in formulating more clearly the questions that needed to be answered; but as to real answers, there were none. It was debated whether the active principle of conception was furnished by the male, as Aristotle had proposed, or whether, as Galen believed, both male and female "seeds" contributed equally to the formation of the new being, since one may have features of both parents. There were discussions about the sequence in which the fetal organs appear and about the manner in which the fetus finds sustenance inside the maternal womb.

All these discussions were interwoven with the theological and astrological considerations prevalent in the Middle Ages. Human physiology was still believed to be subject to distinct influences from the celestial bodies. Constantinus Africanus (c. 1020–1087), a Benedictine monk who labored at Monte Cassino translating medical texts from the Arabic to Latin, discoursed on the position of the fetus and its progressive growth, relating it to the movements of the stars in the sky.[1] Pietro d' Albano (1257–1315), a translator of the works of Galen, expounded on the "informative virtue" present in semen but remarked that it was exerted in strict conformity with the heavens. For medieval medicine, the human body was a microcosm that reflected and epitomized the entire universe.

With the advent of modernity, theories began to be based upon actual observations, but the tools with which to perform these were still primitive. As a result, the hypotheses of even the best minds were often extravagant or far-fetched. René Descartes described human generation as resulting from "the confused admixture of two liquors [male and female genital secretions] acting as a kind of yeast one upon the other, they heat one another." Some "particles" (the male and/or female fluids were supposed to contain elements coming from all parts of the parents' bodies) are actively stirred, "expand, press on the others, and by this means are arrayed in the manner necessary to form the limbs."[2] The heart would also be formed by the aggregation of hypothetical particles set into motion by heat. This heat he likened to "that of wine when it is fermented, or hay when it is stored."

Thus, for Descartes, embryogenesis is akin to fermentation of a particle-containing fluid that is acted on by mechanical forces: heat and movement. It is a mechanist interpretation, as one would expect of the foremost rational philosopher. But it is also overly simplistic and, judged from our present vantage point, childish.

No less erroneous but more imaginative was William Harvey's hypothesis. He proposed that conception is akin to a "contagion" communicated to the womb by semen, much like a magnet communicates magnetism to a piece of iron. He stated that "what makes the seed fertile is analogous to contagion" (*quod facit semen foecundum, analogum contagio*).[3] But this hypothesis strained credibility. That a new human being could emerge from the admixture of two unorganized fluids was unbelievable. Which is why a new hypothesis was fashioned that bypassed the obstacle altogether: the new being was not formed from two liquid seeds but was already fully formed in one of them.

Contrary to (Aristotelian) hypotheses that explained the gradual formation of the embryo from dispersed and amorphous components ("epigenesis"), the new idea saw the embryo as complete from the outset but very tiny. In this hypothesis, called "preformation," there really was no generation, properly so called. There was only growth. The minuscule embryo, or "germ," preexisted whole. All it had to do was expand, and, from its diminutive, invisible state, it would attain the size of a human fetus.

The first formulation of preformation may have been that of the Italian physician and man of letters Giuseppe Aromatari (1586–1660), a respected friend of William Harvey. His idea was expressed in a short work about the generation of plants, only four pages long and written in the form of a letter ("Epistola de generatione plantarum ex seminibus," Venice, 1625). He argued that the adult plant is already present, albeit in miniature, within a small part of the seed. This brief communication made him famous. Personages as illustrious as Marcello Malpighi (1628–1694), the father of microscopic anatomy, and the eminent Dutch naturalist Jan Swammerdam (1637–1680) agreed with him.

Swammerdam, whose observations and beautiful illustrations of insects were not surpassed until the twentieth century, thought he saw in the nymph of the mayfly or fishfly (insect of the order *Ephemera*, whose adult life may be as brief as one day, but whose nymph stage may last months or years) all the structures of the adult form—wings, legs, body segments, and so on—only folded up, compressed, and shortened. Malpighi, in his work on the chick embryo (*De formatione pulli in ovo*, 1672), also attempted to furnish observations that strengthened the preformation hypothesis.

These and other authors believed that the germ was inside the egg of insects and birds. Therefore, it was natural to think that the egg-laying female of the species was principally responsible for procreation. Then the work of Niels Stensen (1638–1686), and principally that of the young Dutch investigator Regnier de Graaf (1641–1673), established that the ovaries of mammals also produce "eggs" (the concept of "cells" did not exist yet). William Harvey and others affirmed that all animals come from eggs, a belief that came to be known as "ovism." For those who adhered to the preformation hypothesis, the preformed embryo existed in the egg. Hence the female, the mother, was preeminent in procreation, and the father had only a secondary or subsidiary role, that of stimulating the embryo's growth.

"Ovism" was contrary to the ancient Galenic idea, since this theory proposed that the fetus somehow formed after the intermingling of male and female secretions. The ovaries were seen in antiquity as "female testes" capable of elaborating a fluid ("female semen"), which commingled with the male semen to initiate the process of generation. But on this not all the ancients agreed. Faulty observation and a hyperactive imagination led some authors to conclude that the female semen was shunted to the urinary bladder. Soranus of Ephesus (c. 98–c. 138), a remarkable physician who practiced in ancient Rome, perpetuated this curious misconception in his influential treatise of gynecology, in which he says that the ovarian secretion is not important in generation, since it goes to the bladder, whence it is eliminated (in the urine) to the outside.[4] Subsequent observers helped to dispel the error but also introduced their own biases. Vesalius, in Book V of his famous *Fabrica,* depicts the correct relations of uterus and fallopian tubes, but the latter are shown as thin, coiled muscular tubes that wrap themselves around the ovaries, much as the epididymis courses over the testis. This shows that the man who made a career of systematically destroying the descriptive anatomic errors and misadvertences of the ancients was still strongly influenced by the idea that the ovaries are "female testes." His disciple Fallopius (Gabriele Fallopio, 1523–1562) gave the correct description of the structures that bear his name.

When Graaf discovered that the human ovary produces eggs, "ovism" was greatly strengthened. Actually, he had not seen the egg cell proper (today's "ovum" or "oocyte"), which is a cell of abundant cytoplasm measuring about 0.1 millimeter in diameter, but that escaped detection with the rudimentary tools then available. He saw a larger structure, the Graafian follicle, a fluid-filled vesicle present near the surface of the ovary, lined by numerous densely clumped, small (follicular) cells; the ovum is embedded in a stubby mass of these cells that projects into the cavity of the vesicle. The true ovum or oocyte was not identified until 1826 by the Estonian-born investigator Karl Ernst von Baer (1792–1876). Still, the "ovist" faction basked in the confidence that followed the ovary-centered discoveries, until a new development fell like a bomb in their midst: Leeuwenhoek's discovery of spermatozoa.

Antonij van Leeuwenhoek (1632–1723) was not the first to observe live spermatozoa, but his masterful use of the simple microscope (provided with one lens, as opposed to the compound microscope, which utilizes a combination of lenses), combined with his vivid descriptions,

made them known to the world.
A straitlaced, coy, and somewhat prudish Protestant, he was
quick to declare that the semen
he examined was not his but
that of a friend afflicted with
gonorrhea. Soon after his observation, it was clear that the
males of all animal species
studied ejaculated fluid containing their own kind of these
fast-moving, tadpolelike living
forms, or "animalcules," as they
were called. Therefore, he did
not hesitate in aligning himself
with the preformationists and

Antonij van Leeuwenhoek (1632–1723).

proposing that the seed was conveyed in these "animalcules." Hence
the well-known drawing by Nicolas Hartsoeker (1656–1725), showing
a tiny infant all balled up inside the head of a spermatozoon. Thus, the
ovist camp, who professed belief in the primacy of woman's role in
generation, was opposed to the "animalculist" persuasion, which
granted this predominance to the male sperm. A heated debate began
between "ovists" and "animalculists" that would last for more than a
hundred years.

However, if a complete, minuscule human being existed inside the
"germ," it followed that this tiny being had its own genital organs, in
which there would be germs containing smaller complete human beings; and these, in turn, carried germs with even tinier human beings;
and so on, generation after generation, indefinitely. Obviously, this
idea ran contrary to common sense. Consider the difference in size between a fully grown man and the head of a spermatozoon. To fit inside
the latter, the little human being had to be less than a millionth the size
of the man. But this difference in size had to apply again between the
little human being and his or her own "germs." And the same is true
for successive generations. So that after only four or five generations,
the disproportion in size between ourselves and the fourth or fifth "en-
capsulated" germ would be unimaginable. To express it, one would
have to write a fraction with a denominator of such huge value as only
astronomers are accustomed to using when discussing the distance be-

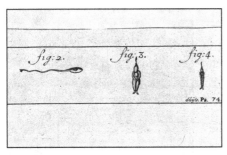

Illustration by Leeuwenhoek showing a compressed, "preformed" human infant inside the head of a spermatozoon. COURTESY OF THE NATIONAL LIBRARY OF MEDICINE

tween the earth and the stars. And this at only the fifth generation!

How could a reasonable person be expected to believe this? In the words of a leading preformationist of the times, if we cannot conceive of such an extreme smallness, "only our mind is to blame, whose weakness we acknowledge every day; but it is no less true that all the animals who have been, are, and shall be, were created at the same time, and all locked inside the first females."[5]

Leeuwenhoek observed many classes of cells, but he was not able to ascertain that all living organisms are composed of cells as primary units. By the same token, he observed the sperm cells pullulating in the ejaculate of male animals of many species, but he failed to understand their role in human generation. A century later, the great Italian biologist Abbot Lazzaro Spallanzani (1729–1799) demonstrated incontrovertibly that in order for fecundation to occur, the male's ejaculate has to come into contact with the female's ovum. Still, he thought the fertilizing power resided in the liquid part of the ejaculate, not in the sperm cells. To him, those tadpolelike "animalcules" were mere adventitious contaminants!

Spallanzani thought the weight of the evidence was in favor of the egg as the site of the germ. He was thus an ovist as well as a preformationist. He devised a series of impeccable experiments that pointed strongly to the role of sperm cells in procreation. Among other things, he demonstrated that once passed through filter paper, semen lost its inseminating potency. And what, if not spermatozoa, could be the elements retained in the paper? But such is the force of our preconceptions that we would ignore the best evidence if it did not corroborate our current beliefs.

No one up to the eighteenth century was prepared to understand the process of generation, but few would admit it. Voltaire was among those few: with his inimitable humor he made fun of the founder of "ovism" (William Harvey, to whom he refers in an essay by the fictional name "Aryvhé"), saying:

> He dissected a thousand quadruped mothers that had received the male's liquor; but after having also examined the eggs of hens, he decided that everything comes from an egg; that the difference between birds and other species is that birds hatch, and the other species don't; [that] a woman is only a white hen in Europe and a black hen in the bottom of Africa. All repeat after Aryvhé: *Everything comes from an egg.*

Lazzaro Spallanzani (1729–1799).
COURTESY OF THE NATIONAL LIBRARY OF MEDICINE

Voltaire ends his exposition with this frank conclusion: "I think we have to resign ourselves to ignorance of our own origin: we are like the Egyptians, who draw so much help from the Nile and do not yet know its source; perhaps we shall discover it one day."[6]

———

Although the microscopists of the seventeenth and eighteenth centuries were unable to solve the enigma of generation, the invention of the microscope is surely one of the major milestones in the history of medicine. Dutch lens makers started producing magnifying glasses of very good quality in the early 1600s, and naturalists quickly used them to study very small objects. Leeuwenhoek was at the forefront of this movement. Although originally a draper from the town of Delft, in western Netherlands, he perfected the polishing of lenses using techniques that he guarded zealously, and he constructed a simple microscope capable of 266-fold magnification.

His microscope, furnished with a single lens, was little more than an improved magnifying glass. But the lens was spherical and very finely polished, and thus kept chromatic aberration to a minimum. It fit snugly into a hole in a flat metallic plate; the specimen was attached to a needlelike holder underneath. The holder was connected to a screw that, when turned, rotated the specimen so it could be examined from all sides. To see the magnified object, the observer had to place his eye very close to the lens.

With this unsophisticated but useful tool, Leeuwenhoek described the microscopic structures of plants, insects, crystals, protozoa in stagnant water, pieces of wood, and many other objects. He was among the first to observe red blood cells (1673), spermatozoa (1677), yeasts (1680), bacteria (1683), and capillaries (1689). He recorded his many spectacular discoveries in Dutch (lacking a university education, he was ignorant of Latin) in hundreds of reports to London's Royal Society, into which he was admitted upon the recommendation of his friend Regnier de Graaf.

Nevertheless, it was a long, arduous road from the discovery of spermatozoa to ascertainment of the role of sperm and ovum in human generation. The physiology of ovulation was unknown for a long time, and was gradually unveiled by a number of observations in several animal species.[7] Graaf's observation of a case of ectopic tubal pregnancy allowed him to infer that the ovum must pass from the ovary to the uterus via the fallopian tube. By the end of the eighteenth century, the Scottish anatomist William Cruikshank (1745–1800) actually observed egg cells in the fallopian tubes of rabbits and inside the uterus, where he was able to see the embryo's implantation;[8] but it was not until the nineteenth

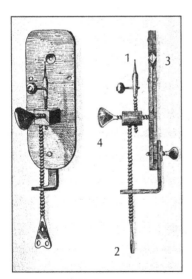

Microscope used by Leeuwenhoek: 1) specimen holder; 2) sample translator; 3) site of magnifying lens; 4) knob for focusing.

century that it was determined, largely through the work of the Swiss scientist Jean-Louis Prévost (1838–1927) and the French chemist-embryologist Jean-Louis-Baptiste-André Dumas (1800–1884), that sperm cells, not seminal fluid, are the fecundating components in sperm.

Throughout most of the nineteenth century, the study of embryology was purely descriptive. The leaders of the field, such as Christian Pander (1794–1865), Martin Heinrich Rathke (1793–1860), and especially Karl Ernst von Baer (1792–1876), made it clear that only extensive microscopic examination could lead to an understanding of embryonic development. With admirable assiduity, they proceeded to do just that; and the result of their untiring efforts was a series of works on the embryologic development of fish, reptiles, and vertebrates in general.[9]

Toward the end of the nineteenth century and beginning of the twentieth, embryology became chemical and "experimental." That is, researchers directed their efforts toward clarifying the biochemistry associated with the morphological changes that their predecessors had described so thoroughly and toward the production of malformations by chemical modifications of the environment. At the end of the first third of the twentieth century, it was still possible, although by no means easy, for a single investigator to master all that was known about the chemistry of embryology. A summary of this knowledge was contained in a treatise of encyclopedic scholarship entitled *Chemical Embryology* (1931), by Joseph Needham (1900–1995).

The study of chemical embryology was furthered by the discovery of "induction," the process whereby the development of one part of the embryo is influenced by the "signals" or "messages" coming from another part. A Nobel Prize was awarded to Hans Spemann (1869–1941) for his work in this area.[10] This opened the way for a new era in embryology, which continues to this day, in which the explosion of knowledge on growth factors, their receptors, and the genetic, DNA-encoded control of their expression is yielding astonishing insights into the enigmatic processes of embryonic life.

When embryology was still a highly theoretical discipline, it was of interest to only a restricted circle of specialists. But, starting in the second half of the twentieth century, the new technology began to be applied to human beings. The effects were like those of an earthquake: its tremors were felt far and wide, and its force proved capable of dislocating the stolid values and supposedly unshakable institutions of society.

In the technique of in vitro fertilization (IVF), the ovum is fertilized by a spermatozoon in a Petri dish or some other laboratory container. Conception takes place under laboratory conditions, outside the body, and independently of the sexual act. Once an embryo is formed, it is implanted in the maternal womb. Robert Edwards (b. 1925), an embryologist at Cambridge, succeeded in fertilizing the egg cell of a woman, Lesley Brown, using the sperm of her husband. The woman delivered the first baby conceived in vitro, Louise Brown (whom the popular press called the "test tube baby"), on July 25, 1978, amid a racket of worldwide publicity.[11] The baby girl was normal and with the passage of time grew to adulthood, conceived, and delivered a normal child herself. The success of this technique was touted as a panacea for infertility. Women with this diagnosis flocked from all over the world to clinics that offered IVF.

It should be noted that in sexual reproduction, millions of spermatozoa actively move to encounter the oocyte. But only one among those millions will actually penetrate the ovum to add to it the complementary genetic material. Is it the one that moves fastest? Is it the one that has the best genetic material to contribute? Or is it pure chance that determines which is the fecundating spermatozoon? The truth, as the eminent scientist Luc Montagnier attested, is that *we do not really know*.[12] No studies exist that could define which of these possibilities holds true. This uncertainty should cool the enthusiasm of those who inject a sperm into the ovum in the technique known as ICSI (intracytoplasmic sperm injection). It will take several generations to find out whether this technological "improvement" has any undesirable effects.

In women who cannot ovulate, fertility can be restored only by using someone else's egg cells. This, of course, blurs the traditional concept of maternity. Formerly, the woman who bore the child was considered the natural mother. Now the woman who donates the ovum can logically claim some legal rights to maternity: after all, the child was conceived with her reproductive cell (ovum) and consequently with half her genes. When donor sperm is used in IVF, similar considerations apply to the father, for the same biological reasons.

The situation became further compounded when "surrogate mothers" entered the picture. Surrogates are women willing to carry the child in their womb, as a service to infertile women incapable of gestating in their own uteruses. When "surrogacy" is done for a price ("uterus for hire"), it lends itself to economic exploitation, just as the commercialization of egg cells has been tarnished by abuse: women who are surrogate mothers, like women who sell their ova for use in IVF, tend to be economically disadvantaged and are pressured by financial need, while the women who use their services tend to be better off. This, of course, is not always the case. But even when surrogacy is done for altruistic reasons, without financial motivation, important social complications may arise. For instance, a woman may decide to act as surrogate mother on behalf of her infertile daughter, as has occurred in widely publicized cases. The baby she carries is both her child and grandchild. The child will grow up with a mother who is not its biological mother, and its relationships with the other close family members are likely to be confusing. This is only one example of how the new technology of reproductive medicine, if uncritically applied, can scramble the lines of filiation that individuals have learned to recognize, and that have been normative of human conduct since the beginning of history.

Furthering the confusion was the advent of cloning. In the technique of cloning by nuclear transfer, the nucleus of an egg cell is removed and replaced with the nucleus of a mature adult cell. The introduction of this mature nucleus is achieved by a sophisticated laboratory procedure that gives a tiny electrical discharge to the recipient egg cell cytoplasm. This not only fosters the penetration of the nucleus but apparently "tricks" the recipient egg cell cytoplasm into believing that a spermatozoon has penetrated. Accordingly, the renucleated cell starts behaving as if true fertilization had occurred, and it starts dividing actively. After a few divisions, the cell is placed inside the uterus of a female suitably prepared to support the gestation, and this "foster mother" then carries the developing fetus to term.

Such was the accomplishment announced on February 23, 1997, by the Scottish biologist Ian Wilmut and his associates. They succeeded in obtaining a newborn sheep, which they named "Dolly," using a cellular nucleus from breast gland tissue of a ewe that had died years before. This was truly a breakthrough. It ran counter to one of the central principles of embryology established since the nineteenth century—

namely, that adult cells are irrevocably committed to the type of specialization that they acquire early in life. Within less than a year, other investigators succeeded in cloning mice; still others followed who, employing the same or similar methods, obtained the birth of pigs, goats, rats, rabbits, horses, and dogs.

An offspring born in this manner is a clone, therefore differing from a normal product of gestation in several important respects. First, it was conceived without a spermatozoon, entirely by laboratory manipulations. Second, its genetic constitution will be identical to that of the donor of the cellular nucleus (i.e., the nucleus introduced to reconstitute the egg cell). Note that this donor cannot be considered to be "the father," since the term "father" is reserved, in conventional speech, for the male that contributes half the genetic endowment to his offspring, while the other half of the genes comes from the mother. In this case only the nuclear donor passes on the totality of his genes, without mixture; hence the offspring is genetically identical to the donor, which is therefore more like the "twin" than the father.

Nothing in the past remotely resembles the puzzles and quandaries raised by contemporary reproductive medicine. In the United States, a National Bioethics Advisory Commission (NBAC) was established in 1995, as soon as it became clear that there was technology ready that could lead to the cloning of human beings. But the truth is that none of the members, although experts in widely diverse fields, had any expertise in the problems they were expected to solve. A full comprehension of these problems requires a reasonable understanding of genetics, molecular biology, and some fundamental concepts of embryology. A working knowledge of selected medical procedures is also essential. But legislators, like lawyers, philosophers, and other experts called upon to form advisory commissions and consulting bodies, often lack this knowledge. Nor can they be expected to acquire it quickly. Becoming even passably familiar with these areas requires a major investment of time and effort.

Hence the need for the public's participation. Fortunately, there is an abundance of sources of information that place these problems into proper perspective and that clearly define the difficulties that must be faced.[13] Hopefully, as a greater segment of the public becomes well informed, a critical mass will be reached that may react with vigor to rash proposals and ill-advised implementation of new technologies and exert pressure to ensure more prudent legislation.

THE HISTORY OF OBSTETRICS

Whether conceived in the laboratory in an asexual manner or by the sexual conjunction of male and female progenitors, a fetus, once reaching maturity, must leave the maternal enclosure. The transit to the outside may be a relatively easy passage or one bristling with potentially lethal problems. It is the goal of obstetrics to prevent such problems.

Obstetrix is Latin for "midwife," and is thought to derive from *obstare,* meaning "to stand before," probably because the attendant stood in front of the woman to receive the baby. For many centuries, childbirth was strictly a woman's affair, and men were systematically excluded from the delivery room. The ancient Romans recognized that women's diseases were different from men's and saw the need for specialists in this area. Medicine was one of the few men's activities available to women in ancient Rome, and many women earned fame and prestige as physicians. Gynecologists and obstetricians tended to be women, though this does not mean that they were restricted to these specialties.[14] Still, the most respected author in the field was a man, Soranus of Ephesus (c. 98–c. 138). Soranus's gynecological advice is full of common sense; he propounded sound hygienic rules that, in a time of clinical slovenliness, must have greatly benefited his patients. His writings formed the basis of a text known as *Moschion* (also spelled *Moscion,* from a Latin writer named Muscio, who paraphrased Soranus), which was highly respected throughout the Middle Ages. It was quoted as an authority with little modification for at least nine centuries.

No new significant advances took place until the Renaissance. Then, with the advent of the printing press, some writings of gyneco-obstetric import began circulating that, curiously, were written by men. One was popularly known as *The Rosengarten,* by a German author, Eucharius Rösslin.[15] First published in 1513, it was well illustrated with woodcuts that reproduced some of Soranus's original material. It was translated into English as *The Birth of Mankind.* Another text that gained notoriety was Jacob Rueff's (1500–1558) *De conceptu et generatione hominis* (1554), also rich in illustrations and harking back to Soranus.

The Renaissance texts berated midwives, accusing them of ignorance and superstition and of following practices that were harmful to their patients. Who could have been the intended audience of those writings? Certainly not male physicians, since they were excluded from attending childbirths. Midwives at the time were taking care of practically all women giving birth. It seems very unlikely that an unsympathetic pamphlet, aggressively derogatory to midwives, might have been written in the supposition that these women would buy it and contribute to its diffusion. Perhaps well-educated rich males, chiefly the aristocracy and clergymen, were the intended public, and the illustrated texts were produced in response to the voyeuristic tendencies of their readers, which these works aimed to satisfy.

Historians have claimed that, with the exception of Britain and her colonies in America, there were practically no males practicing obstetrics in the modern age until at least the year 1800.[16] However, a few men did practice this specialty. This was the case in France, where King Louis XIV, aware that physicians had been acquiring important new skills, permitted one of them, Julien Clément (1650–1729), to deliver a child that the monarch had sired with one of his mistresses, Madame de la Vallière. Others were subsequently summoned to attend the wives of influential aristocrats and even royalty, and an official "accoucheur" was appointed at court. This conferred much respectability to the profession, but it was not a sudden development. François Mauriceau (1637–1709), a skilled surgeon, had published in 1668 his *Treatise of Diseases of Pregnant Women* (*Traité des maladies des femmes grosses*), which went through many editions—at least seven in the eighteenth century. It was translated in England by Hugh Chamberlen (1630–1720) as *The Accomplished Midwife* (1672).

Mauriceau's name is familiar to obstetricians due to a method he devised (known in English-speaking countries as the Mauriceau-Smellie-Veit maneuver) to facilitate the birth of children who present in breech position, instead of the normal "headfirst" during delivery: the obstetrician introduces a finger into the baby's mouth and turns its face posteriorly. Modern textbooks recommend pressing a finger on the baby's maxilla, as this will cause greater flexion of the head and ease the baby's expulsion.

In the past, retention of the after-coming head was a complication frequently fatal to both mother and child. Desperate measures were taken: nothing else could be done. The baby's head was cut open (craniotomy) and the brain destroyed: only then did the soft, membranous bones of the skull collapse, and the head of the dead baby was extracted. When the baby presented in transverse lie (i.e., its major bodily axis lying perpendicular to the maternal body's longitudinal axis), the horrible method of dismembering was the only option available. There was no anesthesia, and cesarean section was almost uniformly fatal. Thus, piece-by-piece removal of the fetus was resorted to, with forbidding-looking instruments appropriate for craniotomy and fetotomy; in Roman antiquity, even a "decapitator" was devised to facilitate the separation of the head.

All this sounds terribly cruel and barbaric, but one must remember that these gruesome measures were resorted to after the struggle of delivery had been very protracted and when the fetus was believed to be dead. Physicians were usually careful to secure the permission of the religious authorities before going ahead with the ghastly procedures, to avoid being accused of intentional wrongdoing. However, there were instances when the mother's death seemed imminent, and in those cases an influential obstetrician advised to "treat the baby *as if* dead."

One historically important development in delivery was the invention of the obstetric forceps. Although different versions of this instrument existed in ancient times, the modern version is commonly traced to the Chamberlen (originally Chamberlain) family in England. William Chamberlen was a French Huguenot who settled in

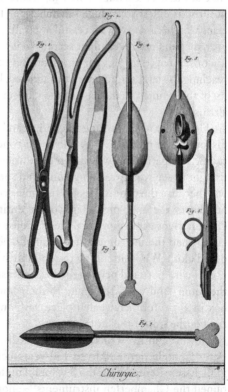

Obstetrical instruments portrayed in a 1772 textbook edited by Denis Diderot. Their forbidding appearance gives an idea of the radical and, by necessity, cruel manner of their use. COURTESY OF THE NATIONAL LIBRARY OF MEDICINE

England in 1569. One of his sons, Peter Chamberlen (1560–1631), had a brilliant career as a surgeon and obstetrician and attended the wives of James I and Charles I in childbirth. However, he was compromised in a failed antigovernment rebellion, arrested, and then exiled to Holland, where he died. It was probably Peter Chamberlen who developed the obstetric forceps. The exact timing of his invention is unknown, due to the secrecy with which this instrument was surrounded.

The idea may have come from the use of large spoons and other tools that midwives were known to use in order to ease the slipping out of the infant's head, much as a shoehorn is used to facilitate the slipping on of a shoe at the heel. Chamberlen's son, also named Peter (1601–1683), and in turn his eldest son, Hugh Chamberlen (1630–1720), were all physicians who became very successful at treating difficult childbirths. However, they did not reveal their secret. Today, most people would think it unconscionable to prevent an invention of benefit to mankind from reaching the largest number of the afflicted, but this was not the mentality of those times. In the case of the Chamberlens, greed prevailed over altruism.

They were always trying to confuse and disorient those who might discover their method. They allowed no one with them in the delivery room. The parturient woman, in the throes of her travail, was too distraught to see what was being done; and she could not have seen, anyway, because the prudishness of the times concealed everything under bedsheets: the male physician worked entirely covered by these. The Chamberlens came into the patient's house carrying a large box, much larger than the instrument it contained and covered with an embroidered cloth. While attending the childbirth, they produced distracting noises with pots and pans to further confuse those who stayed outside the room. And upon returning to their home, they kept the precious box in a carefully concealed place, fashioned for this purpose, under the floorboards of the attic.

However, in 1813, this hideaway was discovered at the house where Peter Chamberlen had lived 130 years before: Woodham Mortimer Hall, Essex. Remarkably, the box found under the floorboards still contained the Chamberlen instruments[17]—not only the forceps, but two other interesting devices. One was a vectis, a single metal blade, somewhat curved at one end, that was tucked under the baby's nape, and then traction was exerted on the handle at the other end. The other was a fillet, a long, ribbonlike segment of a pliable material, silk or leather, provided with a handle and meant to be looped around the baby's head; then the handle was pulled to expedite the delivery. These inventions are usually ignored in textbooks on the history of medicine, because the only instrument that became widely adopted was the obstetric forceps.

The Chamberlens' secret eventually came out. The obstetric forceps was enthusiastically adopted by many physicians, although its use was controversial from the outset. A number of variants were devised. William Smellie (1697–1763) added a "pelvic curvature" to the blades and tried to make them of materials other than metal, "to spare the mother the chilling clink of interlocking steel as he set to work." Smellie was the biggest name in European obstetrics during the eighteenth century and the author of *Treatise on the Theory and Practice of Midwifery* (1752–64). For all his fame, his rough manner made him unpopular. "A great horse godmother of a he-midwife," he was famously called by an irate midwife.

His successor and sometime pupil was William Hunter (1718–1783), older brother of the renowned surgeon John. William Hunter had all the charm and winsomeness that Smellie, his mentor, lacked.[18] His social talents earned him many students and made it acceptable for men to practice obstetrics. He is reputed to have toasted to his friends—who included the literati Tobias Smollett (1721–1771), James Boswell (1740–1795), and others—"May no English nobleman venture out of this world without a Scottish physician, as I am sure there are none who venture in," thereby expressing his confidence that male accoucheurs in his time were bringing the newly born into the light of day.

William Hunter's masterwork was *The Anatomy of the Human Gravid Uterus*. This book, like Vesalius's *Fabrica*, combined expository strength and exceptional beauty of illustrations. Hunter's book benefited from the work of the Dutch artist Jan van Rymsdyk (c. 1730–1788 or 1789), whose superb drawings are masterpieces in themselves.

Yet male obstetricians were opposed for several reasons: it was deemed immodest for a male physician to take care of women in childbirth; midwives resented men encroaching on their turf; and there was widespread hostility to the use of instruments. Since male physicians were initially called in only to treat complicated childbirths, they were likelier than midwives to resort to the use of instruments. Midwives claimed that only the hand, and especially the female hand, had the softness and the conformation required to effect an easy delivery. In-

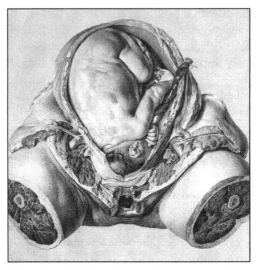

Illustration for William Hunter's The Anatomy of the Human Gravid Uterus *by the eighteenth-century Dutch artist and medical illustrator Jan van Rymsdyk. While acknowledging the superior draftsmanship and aesthetic value of this artist's work, some critics have discerned a dehumanizing element in his depiction of sectioned bodies.*

struments were intimidating and often harmful. Improper application and excessive force resulted in trauma to the baby, the mother, or both. William Hunter himself declared that, contrary to most surgical interventions, "the *Forceps* (midwifery instruments in general, I fear) upon the whole, have done more harm than good."[19]

The controversy lasts to this day. As often happens in medicine, what was initially hailed as a boon was later discovered to be dangerous, and, in the end, a compromise had to be reached: properly used, and recognizing its limitations, it may be of benefit in selected cases. However, in recent years, largely because of the tendency to resort to elective cesarean section, few physicians are being properly trained in the use of forceps, and the instrument may soon be completely abandoned.

Instrumental deliveries were refined with the creation of the vacuum extractor, invented in the 1950s by Tage Malmström of Gothenburg, Sweden (although the thought of applying a "suctorial force" to

A cartoon depicting a "man-midwife," in criticism of the practice of midwifery by men, from the frontispiece of John Blunt's book Man-Midwifery Dissected, or, the Obstetric Family-Instructor for the Use of Married Couples and Single Adults of Both Sexes *(London: Samuel William Fores, 1793). Note the allusion to the use of hard instruments by the male, not the female, part of the man-midwife.* COURTESY OF THE NATIONAL LIBRARY OF MEDICINE

aid in the extraction of the baby had been mentioned by James Simpson in the nineteenth century). A "suction cup" or ventouse of appropriately soft material is placed on the presenting head of the baby, and traction is exerted to facilitate its progression to the outside. As in the case of the forceps, considerable skill and experience are needed to apply it properly. Incorrect placement, undue traction, misapplied vacuum, and wrong manipulation can cause traumatic lesions to mother and child. Like the forceps, its use has become less frequent as the problems it purports to solve are approached by elective cesarean section.

The cesarean section, or delivery of the young through incision of the mother's abdomen, is one of the major developments in the history of medicine. Its name is thought to derive from the legendary account that Julius Caesar was brought into the world in this fashion, yet there is no historical evidence that this was the case. In fact, the evidence argues against it, for in ancient times, before the invention of anesthesia and antimicrobials, no woman subjected to this surgical operation survived, and it is known that Julius Caesar's mother was alive years after his birth. Some think that the term "cesarean" comes from the Latin *caedere,* "to cut" (past participle *caesus,* "cut"). The Romans believed that if a pregnant woman died, she must not be buried without first being cut open, in the hope of salvaging her child; if the child was found dead, it had to be buried separately. These facts were consigned in the law first known as *lex regia,* proclaimed under King Numa Pompilius in approximately 715 B.C. and later named *lex caesarea,* to flatter the first Caesar.[20]

Most stories about successful cesarean sections in the preanesthetic era are probably untrue. For example, there is the oft-repeated story that Jacob Nufer, a Swiss sow gelder, operated on his wife in 1500, with permission of the authorities. Even earlier, Guy de Chauliac (c. 1300–1368), one of the greatest surgeons of his time, wrote in his textbook of surgery about the advisability of incising the left side of the maternal abdomen—not the right, where there is danger of cutting the liver—in order to extract a baby by cesarean section.[21] Although this advice shows anatomical awareness, he also counseled maintaining the mother's mouth and vaginal orifice open, so that the baby would be able to breathe—a precaution that, as we now know, is sheer nonsense. His recommendations were repeated by succeeding generations. Ambroise Paré, in the sixteenth century, realized that there was no way the recommended measures would help the child to breathe.

Paré opposed the performance of cesarean sections on living women. Having heard reports of "women who had their abdomens cut open, not once but several times," he was prompted to declare that such stories were false, for the reported fact was "impossible." The hemorrhage provoked by this intervention would be fatal, and "the scar would prevent the uterus from ever dilating again." In sum, this procedure, in his estimation, was "too dangerous and offered no hope."[22]

It was not until late in the eighteenth century and the beginning of the nineteenth that credible reports of successful cesarean sections began appearing from various countries. The first cesarean section in the British Empire is attributed to James Miranda Barry (c. 1795–1865), a military surgeon who traveled widely with the British army and who turned out to be a woman who had masqueraded as a man for more than forty years. The scandalous fact of her real gender was not revealed until she was laid out for her funeral in 1865. Her life has been the subject of books, plays, and movies. In North America, it is claimed that the first cesarean section was done by Dr. John Richmond in Ohio in 1827. The early operations were still done without anesthesia, and the maternal mortality was extremely high. However, as anesthesia, asepsis, and technical improvements became available, the outcomes improved dramatically.

Gradually, the complications of parturition yielded to medical know-how. Excessive hemorrhage was efficaciously combated with ergot; pain, with anesthetics; prolonged labor, with forceps. The dreaded, highly lethal "childbed fever" or puerperal fever, now known to be caused by bacteria, mainly streptococci, was routed with simple measures of hygiene. Without any knowledge of bacteriology, Alexander Gordon (1752–1799) and Oliver Wendell Holmes (1809–1894) independently suggested that the disease was spread from patient to patient by midwives and physicians and not necessarily through "miasmata," or fetid airs, as was widely believed. Next, the Hungarian physician Ignaz Semmelweis (1818–1865) observed that, among hospitalized pregnant women, those cared for by doctors fell victim to puerperal fever more often than those attended by midwives. After extensive research, Semmelweis concluded that the doctors' hospital routine was the reason for this phenomenon. Their day started early in the morning with attendance in the morgue, where they were expected to perform autopsies. Following this, without transition, they made their rounds in the obstetrics ward, and here they performed gynecological examinations. The inference seemed inescapable: the physicians were communicating the disease to the women in labor. Semmelweis reinforced his hypothesis when a friend of his died from an infection contracted through an accidental skin cut while performing an autopsy. The pattern and structure of the lesions revealed at this friend's autopsy were strikingly similar to those of women dead of childbed fever.

By the simple method of enforcing scrupulous hand washing with chlorinated water, the frequency of childbed fever fell precipitously. Semmelweis corroborated his theory, showing that, in medicine, ignorance of the cause of a disease does not always preclude the effectiveness of its treatment. But the mere thought that physicians were the actual transmitters of the disease, and that their ignorance and recalcitrance had inflicted so much suffering upon pregnant women, was more than the doctors could bear. In a dark page of medical history, they reacted rancorously against Semmelweis, whose ideas they ridiculed and vilified. He was fired from his post and driven out of Vienna. Years later, he would die a broken man, in an asylum for the insane, apparently after being savagely beaten by the asylum personnel. It would be close to half a century before his ideas gained favorable reception and universal application.

Semmelweis's tragic life has tempted biographers, among them the physician-writer Sherwin B. Nuland,[23] who concluded that Semmelweis was not just a martyred victim of incomprehension and prejudice but also an insecure man, whose sense of alienation in Vienna (where he always saw himself as a foreigner) made him overreact to criticism. A more restrained reaction, and a little more perseverance in defending his ideas, claims Nuland, might have earned Semmelweis the recognition he deserved.

Ignaz Semmelweis (1818–1865).
COURTESY OF THE NATIONAL LIBRARY
OF MEDICINE

With the virtual elimination of puerperal fever, childbirth became much safer. But the otherwise undeniable obstetrical advances also had undesirable effects, chief among them abuse of the surgical procedure. The current literature asserts that cesarean sections are associated with lower maternal mortality rates than vaginal deliveries.[24] As a result, many women now ask to deliver by cesarean section, and a considerable proportion of physicians readily acquiesce to the demand, irrespective of the medical indications. An inquest showed that 69 percent of obstetricians in the United Kingdom and 50 percent in Israel would respond positively to such a request.[25] The frequency of elective cesarean sections is increasing: it was 40 percent in the United States in 1980. In 2005, one estimate was 30.2 percent, but this still represented a 46 percent increase since 1996. In Rio de Janeiro it approaches 90 percent, and in England, where it was only 3 percent in 1950, it had risen to 20 percent by 2000,[26] and is still rising. In 1989, in some parts of Canada, one out of every four deliveries took place by cesarean section.[27]

FINAL CONSIDERATIONS AND CONCLUSIONS

There is no denying that childbirth, an inherently risky biological phenomenon, is much safer for present-day women than it was for their grandmothers. This safety is taken for granted in industrialized countries. For instance, in Britain in 1935, the maternal mortality rate was 400 per 100,000 births; at present it is only 11.4. Puerperal fever has been virtually eliminated. Disastrous deliveries, in which mother and child either succumbed or survived with ghastly tears or lifelong invalidism, are largely a thing of the past. Prolonged, difficult parturitions, whose appalling devastation used to be accepted with fatalism, are now largely prevented by prenatal surveillance and planned cesarean section; and, as pointed out, some experts deem the risk of this intervention as not higher, and possibly lower, than that of vaginal delivery.

However, all is not well. Though the professed goal of obstetrics is to satisfy the needs of women in all that pertains to pregnancy and delivery, many women have raised their voices in recent years to deny that this goal has been met. In the industrialized world, they see the "medicalization" of childbirth as a wrong done to this quintessentially female experience.

Some women claim that to be born in cold, impersonal hospital conditions, utterly devoid of human emotional content, is not to be born at all; it is more like being violently dragged into existence, or entering the world as did Macduff in Shakespeare's *Macbeth*: "...from his mother's womb / Untimely ripp'd" (act 4, scene 1).

Attempts to "rescue" childbirth from the impersonal, dehumanizing state to which official medicine had reduced it were originally spearheaded by men. Some physicians, such as Fédéric Leboyer and Michel Odent, inspired by the Italian educator Maria Montessori (1870–1952), held that the very first impressions of a baby are crucial for its later development and suitably modified the delivery suite with dimmed lights, soft music, agreeable scents, a warm bath for the mother, and complete freedom to adopt any position that she preferred during delivery. All this, of course, came at a price that many mothers could not afford.

Grantly Dick-Read (1890–1959) held that the pain of childbirth is largely the result of fright and the tension created by doctors' rituals and the manipulations of hospital personnel. He devised a method of "natural birth control," which he described in books that were bestsellers and translated into many languages. Of the same persuasion was the French gynecologist Fernand Lamaze (1891–1957), who gained notoriety with his method of psychological preparation and controlled breathing to reduce the pain and stress of childbirth.

These would-be reformers themselves became the target of feminist criticism. The Lamaze method, although springing from good intentions and unimpeachable pedagogical motives, seemed couched in the paternalistic pronouncements of an intolerant patriarchy. If a woman experienced much pain, she was blamed for her inability to learn how to relax. The responsibility rested entirely on her shoulders, and if the results were not as expected, the implication was that she was incapable of discarding her ignorance and prejudices. Hence the "back to home childbirths" movement.

However, where obstetrics truly has failed to meet its responsibilities is in the care of women in poor countries. It is common knowledge, and widely reported in the press, that in sub-Saharan Africa, women face a 1-in-16 chance of dying in childbirth. Deaths due to pregnancy and delivery surpass in frequency those due to AIDS. The number of maternal deaths per 100,000 live births is 2,000 in Sierra Leone, 1,900 in Afghanistan, 1,800 in Malawi, and 1,700 in Angola, and only 12 in industrialized countries. Infant mortality, of course, is also extremely high in sub-Saharan Africa: 100 per 1,000 births, compared to 34 in Southeast Asia, 30 in Latin America, and about 6 in industrialized nations. At the close of the twentieth century, 515,000 women died yearly in childbirth throughout the world. Of these, close to half were in sub-Saharan Africa, 10 percent in Southeast Asia, 6 percent in North Africa and the Middle East, 4 percent in Latin America, and only 1 percent in industrialized countries.[28]

Certainly, these statistics cannot be entirely blamed on the medical profession. The problem is largely political, economic, and cultural. However, physicians bear some responsibility. Alexandre Minkowski (1915–2004), a medical expert in maternofetal diseases who worked in many destitute areas of the world, concluded that the education of midwives would immensely improve the dismal conditions that he witnessed. Midwives, he pointed out, are close to the population and know very well the traditions and customs of the region in which they work. Local ways must be respected if better sanitary conditions are to be implemented, and experience shows that midwives, emerging from the ranks of the common people, are uniquely qualified to do this. In rural areas, they communicate and empathize with the patient, even when they come from another country, better than do physicians. Without the cooperation of local midwives, the best government plans for improved sanitation end in failure. But in spite of these observations, Minkowski's recommendations have been met with the stubborn resistance of organized medical societies. At least in some Arab countries, the doctors' associations have adamantly refused to let the care of pregnant women pass into the hands of midwives.[29]

5

PESTILENCE
AND MANKIND

*Brief Overview of the History
of Some Infectious Diseases*

Human civilization has been repeatedly and profoundly influenced by recurrent epidemics. In the interest of brevity, the following comments refer to only a few historically important epidemic infectious diseases.

PLAGUE

In antiquity, the plague of Athens (c. 430–425 B.C.) was well documented. According to Thucydides, the historian who witnessed it, 300,000 Athenians—one of every three—died. The patients had high fever, hiccups, bilious vomiting, intestinal ulceration, and diarrhea, swiftly followed by death. From this description, several interpretations are possible. The retrospective diagnosis of bubonic plague has been made somewhat laxly, as it does not quite jibe with the facts that Thucydides described. Bubonic plague is characterized by high fever, chills, malaise, headache, prostration, and the painful swelling and suppuration of lymph nodes, particularly those of the inguinal region and axillae (lesions popularly called "buboes," hence the name bubonic plague). The Athenian epidemic has also been attributed to influenza, measles, and typhus, and, more recently, to the Ebola virus. Unfortunately for medical history, the ancient Greeks cremated their dead and there is thus no well-preserved archeological material to which the modern techniques of molecular biology might be applied.

It is otherwise with the Black Death, the disease that, starting with the pandemic of 1348–1351, cut great swaths of destruction through medieval Europe. The epidemic came in waves, every twenty-one to twenty-five years. At least for the second wave, DNA analysis of mortal remnants has proven incontrovertibly that the cause was bubonic plague.[1] The disease caused unimaginable devastation. In England alone, it is estimated that 45 percent—and across Europe about one third—of the population died. Small towns were entirely annihilated. In large cities, few survivors lived to witness the harrowing suffering of the community. Bread and other foodstuffs were hard to get when bakers and land laborers had died and the fields lay fallow. In the towns, people literally fell

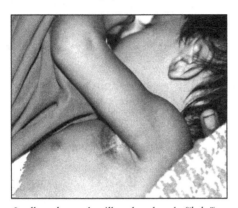

Swollen, abscessed axillary lymph node ("bubo"), a characteristic feature of plague.

dead on the streets, and, there being no one left to bury them, the bodies rotted, abandoned. The survivors were obligated to dump cadavers into huge mass pits, unceremoniously (the priests were dead or had fled to escape danger) and in great haste, up to five or more layers deep. As the disease progressed, those who could ran away, hoping to outrun the plague. The social effects were terrifying: families were decimated, as the survivors often felt the need to leave the sick behind in order to stay alive. The traditional social hierarchies were disrupted: the disease affected the rich and poor, the powerful and weak, alike. Domestic animals died on the farms for want of care. Orphan children wandered alone. Everywhere the spectacle was one of ruin, desolation, and death. Worst of all, these awful tragedies repeated themselves with every new outbreak, for several centuries.

It was not until the nineteenth century that the cause of this disease was discovered by Alexandre Yersin (1863–1943), a Swiss-French physician, who was studying plague in Hong Kong. He detected a bacterium abundantly present in the swollen, abscessed nodes of plague patients. When mice were inoculated, the disease was reproduced. Yersin had a strong background in microbiology, having been a member of the brilliant group of bacteriologists under Pasteur. Impelled by his adventurous spirit, the pull of wanderlust, and, some say, in order to escape the asphyxiating weight of his illustrious mentor's ego, Yersin signed up as a ship's doctor and sailed to the Far East. He became intrigued by, then enamored of, Asian culture as it existed in Vietnam, at the time a part of French Indochina; he spent the rest of his life there. Clearly an idealist and a free spirit, he devoted the remainder of his years to improving the health of the local population. Despite being a citizen of the imperialist nation that oppressed them, he was held in high esteem by the colonials for his altruism, self-

The bishop of Marseille, in 1720, amid the plague-stricken.

abnegation, and sincere sympathy. A hospital in Vietnam bears his name, and his tomb is venerated to this day. He is one of the few European physicians whose memory is held in high regard.

Discovery of the bacterial origin of plague was made more or less simultaneously by the Japanese investigator Shibasaburo Kitasato (1852–1931), who had trained in Germany. However, some of his statements were confusing or contradictory, so the bacterium ended up being called *Yersinia pestis*, in honor of Yersin. The bacterial DNA is similar to that of *Escherichia coli*, which places the Yersinia group in the family of the intestine-inhabiting enterobacteria.

Once the cause was known, the manner of transmission could be determined. It was learned that rats are infected first, passing the disease on to parasite fleas. *Xenopsylla cheopis,* the Eastern rat flea, sucks the bacteria-containing blood of a rat, then passes the disease on to a human being. Other rodents and even cats and dogs may be infected, but the rat flea is by far the principal vector. Various peoples realized that rats begin to die just before the outbreak of a plague epidemic. In the Yunan province of China, inhabitants of rural hamlets were known to abandon their homes as soon as they noted an unusual number of dead rats lying about. Today, it is clear that the elimination of plague depends on abolishing the circumstances that favor the proliferation of rats close to human habitation and the passing of fleas from rats to human beings. These conditions include the accumulation of unprotected refuse and crowded dwellings in and around rat-infested sites. An outbreak of plague is an indictment of deficient health standards at the site of the epidemic.

Knowledge of the cause and manner of transmission of plague makes its prevention seem simple and straightforward. But before such knowledge was available, many believed that plague was caused by venomous bad odors or "miasmal" emanations. According to this notion, poisonous effluvia conveyed invisible, harmful particles or "atoms." They could come from stagnant air near marshes, from caves underground, from rotting corpses, from trash, and from many other origins. Stench was the perceptible manifestation of the noxious quality of the air. This has been called the "miasmatic theory," and it is amazing how well it agreed with all the factual observations.

Plague was more common in the summer. This was logical, since disagreeable odors were then more frequent. Before the invention of modern sewers, filth and muck ran in open ditches, and people used to throw their refuse through the window into the street or dump it in back alleys. Before the invention of automobiles, the horse was the principal means of transportation, and there were large piles of manure on every street. Before there were paved streets, the public byways were ankle deep in mud and puddles in the rainy season. And before there was a modern system for collection and disposal of trash, large mounds of garbage rotted in the open. Today we know that these conditions favor the reproduction of rats and thus increase the likelihood of transmission of disease to humans. But all these observations could fit just as nicely with the "miasmatic theory."

By the same token, it was argued that the wearing of thick fabrics or shaggy clothes favored contagion by trapping noxious emanations. Was it not true that perfume could remain for a long time in well-kept garments? Doctors who visited patients wore smooth, waxed robes to impede the harmful effluvia from attaching to their clothing and a mask with a beaklike protrusion in which they packed odoriferous plants or perfumed towels, presumably to counteract the potentially harmful stenches they could encounter. Today we know that fleas find it easier to colonize a thick or shaggy fabric than a smooth, hard surface. Likewise, people knew that practitioners of certain trades, such as wool merchants, were likelier to get sick than were marble or glass workers. The reason is now clear to us: fleas would rather settle in wool than in glass. But this, too, was perfectly explainable under the "miasmatic theory." Indeed, the congruence of the observations with the miasmatic theory had a perverse quality to it,[2] as if it had been designed especially to deceive mankind.

With appropriate preventive measures and the invention of antibiotics, plague is easily controlled. From its former apocalyptic proportions, it has now been reduced to between 1,000 and 2,000 cases worldwide per year. Still, the U.S. Centers for Disease Control and Prevention (CDC) warns that to speak of "eradication" is ill advised. The disease persists in animals. Occasional epizootic outbreaks are discovered, and the possibility of a comeback of this scourge cannot be ruled out.

LEPROSY

Egyptologists have reason to believe that leprosy existed as far back as 2400 B.C. The Ebers Papyrus contains references to lesions that may have been lepromatous. Retrospective diagnosis, however, is fraught with uncertainty, particularly when it adverts to remote eras in which people were unconcerned with classifications and apt to view things in ways to which we are complete strangers. Cutaneous rashes or ulcers of varied causes may have been given the same name. Still, some lesions are fairly characteristic, and this lends credence to the belief that an ancient Indian medical text, the *Sushruta samhita,* dated 600 B.C., provides descriptions of leprosy. Likewise, archeological evidence contributes to the diagnosis of ancient diseases, as some diseases leave their unmistakable imprints on the skeleton. Leprosy's lesions of the skull, together with the characteristic swollen lips and subcutaneous nodules that show as embossments in the patient's face, bring about a resemblance to a lion's head (so-called *facies leontina*). These pathologic changes are so characteristic that traditionally they have been thought to suffice in establishing a diagnosis. In technical jargon, they are "pathognomonic."

Ashen-colored or hypopigmented skin patches; bulging lumps dotting the cutis; erosion of facial bones, especially of the nasal bridge, imparting to the nose a sunken, crushed appearance; infected ulcers, wounds, and mutilations, the result of neglect (the patient ignores them, due to the insensitivity to pain that is a feature of the disease)—these are the repugnant features of the leprosy patient's aspect. Lepers were therefore ostracized. A Babylonian omen text says that a man so disfigured "has been rejected by his gods and is to be rejected by mankind." In China, there had been similar exiles at least since the time of Confucius. In a well-known passage of Book VI of the "Conversations" (*Luan yü*, the Confucian opus commonly known in the English-speaking world as the *Analects*), Bo-niu, a favorite disciple of the master, is afflicted with a disfiguring disease that commentators have identified as leprosy. Wishing to hide from his neighbors, he secludes himself in his house, where Confucius goes to visit him and tenderly holds his hand from outside the window. Next, the sage ruefully exclaims, "Oh, that such a man should have contracted such a disease!"[3]

During the European Middle Ages, discrimination against lepers was at its worst. They could not marry, touch infants or young people, keep the company of anyone except other lepers, wash or bathe in public streams, or live outside special compounds, the leprosaria or "lazarettos" (so named after Saint Lazarus, probably a leper himself, who begged at the door of "a rich man who dressed in purple and feasted every day" [Luke 16:19–31] and is not to be confused with the Lazarus resuscitated by Jesus Christ). Lepers were forced to wear clothes that made them recognizable and to sound a bell warning others of their approach. These unfortunate patients were, in a sense, "dead in life," and so there were actual funeral services held for them, a sign that, as far as society was concerned, they were no longer members of the community of the living.

Starting in the eleventh century, the disease spread, peaking in the thirteenth century. Whereas there had once been just a few hundred leprosaria in Europe, there were now about nineteen thousand. Then, in the fourteenth century, for reasons that no one has understood, the number affected began to decline. This lowering frequency long antedated the advent of effective therapy, which did not become widely available until the twentieth century.

The discovery of the bacillus causing the disease, *Mycobacterium leprae*, was made by the Norwegian physician Armauer Hansen (1841–1912) in 1871, working in his native Bergen, where 2.5 percent of the population was infected. The first treatment of some benefit was the administration of the oil of chaulmoogra seeds, from a tree of the genus *Hydnocarpus*. This treatment, made known to the Western world by the British physician Frederic John Mouat in 1854, had been part of the traditional medicine of Asiatic populations for many years. In India, legend has it that Prince Rama, afflicted with leprosy, retired into the woods to meditate, feeding on fruits and vegetables. He tasted chaulmoogra nuts, liked them, and was cured. A similar legend flourished in Burma. However, this therapy had serious drawbacks: the benefit was inconsistent, orally administered chaulmoogra oil caused severe, uncontrollable nausea, and injections were quite painful.[4]

In the twentieth century, the therapeutic arsenal was enriched by effective drugs. Promine and dapsone, available in 1940, are sulfones acting as bacteriostatics—that is, they inhibit bacterial growth by interfering with the bacterium's capacity to synthesize folic acid, essential for its proliferation. The advent of rifampicin brought in a truly bactericidal agent. With today's multidrug therapy, real cures are achievable. In 1982, the World Health Organization (WHO) reported the alarming figure of 10 million to 12 million persons affected by leprosy, chiefly in Southeast Asia, China, India, and some parts of Latin America. Since then, the massive availability of the medicaments has wrought dramatic improvements. Starting in 2001, the number of reported new cases has been falling at the rate of 20 percent annually, and in 2004 WHO estimated the number of diseased persons at 286,000. In the more developed nations, this disease no longer exists in endemic form; a small number of cases are discovered yearly, almost all imported from less well protected areas of the world.

Syphilis

More than leprosy, syphilis exemplifies how a disease can explode into devastating epidemics created by social upheavals, wars, invasions, changing sexual mores, migrations, and other human activities. It has been said that "civilization and syphilization have advanced together in the world." Historically, this disease caused greater harm than the much-chronicled medieval plagues because it found two powerful allies: bigotry and fanaticism. Due to the thick veil of pruriency with which many moralists concealed all that pertains to sex, epidemics of syphilis have continued for the last five centuries.

Some scholars believe that syphilis was imported from America to Europe. This claim was based on the alleged lack of paleopathological evidence for the existence of the disease in Europe before the colonization of the New World. This idea did not go unchallenged, and at present the matter seems far from settled. Dissenters feel that syphilis had long existed in Europe but went unrecognized and that European skeletons with telltale lesions do exist. Others contend that the causative agent—which we now know is a 12- to 20-micrometer-long, 0.10- to 0.18-micrometer-wide, corkscrew-shaped bacterium or spirochete of the *Treponema* group—underwent a mutation.

Indeed, diseases caused by other bacteria of the same group, such as yaws, bejel, and pinta, collectively called "treponematoses," are very similar to syphilis and have long existed in tropical America. The genome of these bacteria closely resembles that of the syphilis-causing microorganism. Therefore, it is also possible that one of the bacteria causing the treponematoses was imported to Europe, where the different environmental conditions forced it to adapt and to acquire characteristics that made it more virulent. For instance, it could have gone from living in the patient's skin in tropical climes and being transmitted by direct skin contact to residing in the mucosa of the genital tract in colder latitudes and being transmitted only by venereal contact. In addition, the possibility of two bacterial species merging to create a new strain should not be excluded.

Before the venereal mode of transmission was understood, contagion was attributed to a variety of causes that today seem absurd. Syphilis was thought to be due to wearing linen undergarments, which had begun to replace wool and leather; to poisoning, thereby justifying the disease in kings and potentates; to unfavorable weather; or to unwholesome, fetid air: Cardinal Wolsey was accused of giving syphilis to King Henry VIII by whispering in his ear. Though the king suffered from this disease, we can be sure his manner of acquiring it was altogether different.

The prudish European nineteenth century produced its own bigoted hypotheses. Julius Rosenbaum (1807–1874), a learned and prestigious physician of Halle, maintained that syphilis was directly caused by moral laxity and debauchery. He made this claim even as Pasteur, Koch, and a host of other brilliant bacteriologists were preparing to revolutionize our understanding of infectious diseases. Rosenbaum's hypothesis was that depravity caused venereal diseases and that, if the genital organs were used for procreation only, as dictated by divine and natural laws, these maladies would not exist at all. It was in their use beyond their natural goal, what he deemed the abuse of sexual organs for pleasure, that lay the real cause of genital pathology. Having thus set forth his hypothesis, it was necessary to understand all aspects of debauchery. With this justification, the erudite doctor embarked on a description of every conceivable sexual practice,[5] protesting hypocritically that he was forced to examine, for the sake of medical knowledge, the loathsome aspects of depravity. But all the time he had trouble concealing the delight that he derived from his investigation.

While venereal transmission was ascertained, the distinction between syphilis (also called "lues," from the Latin for plague) and gonorrhea had yet to be established. Many thought the two conditions were one and the same. The famous Scottish surgeon John Hunter believed that it was impossible for two different diseases to exist simultaneously in the same organ or tissue. In an episode that is the stuff of legend, he inoculated himself with the vaginal purulent secretion of a prostitute affected with gonorrhea by puncturing his own penis with a lancet previously immersed in that fluid. He developed syphilis! Of course, his theory was wrong: it *is* possible to have more than one venereal disease at the same time. (Today, when physicians can easily cure the early lesions of syphilis, the concern is that the genital ulcers may facilitate the entry of the AIDS virus, which causes a much more serious condition.) The authenticity of the anecdote has been questioned, but some think that his heroic or rash autoinoculation explains why Hunter postponed his marriage to Anne Home.

A long list of illustrious personages of history were ravaged by the spirochete. During the Renaissance, patients included King François I of France (1494–1547), whose mother said "he has been punished where he sinned"; King Henri III (1551–1589) of France, noted for combating the Huguenots; the great Italian artist Benvenuto Cellini (1500–1571); Pope Alexander VI (1431–1503), of the infamous Borgia family; and his son, Cesare Borgia (c. 1475/76–1507), duke and captain of the army, who, it is said, refused to give audience in order to hide the luetic lesions that disfigured his face. Pope Julius II (1443–1513), famous as a patron of the arts and friend of Michelangelo, would not allow his feet to be washed as prescribed in a Catholic ritual, so as to hide the luetic lesions that covered them. In Russia, Tsar Ivan the Terrible (1530–1584) is thought to have been induced to commit acts of revolting cruelty on account of the syphilitic cerebral involvement that rendered him mad. The Tudors of England, it is generally believed, were often luetic. Besides Henry VIII, who fathered a series of stillborn infants (strong presumptive evidence of lues) and complained of suspicious ulcerations of the thigh, his son Edward VI (1537–1553), a sickly youth, may have died of a combination of tuberculosis and syphilis.

Since the Renaissance there has been firm evidence of repeated epidemics of syphilis and other sexually transmitted diseases. Syphilis was clinically distinguished from gonorrhea in 1873 by the work of Philippe Ricord (1800–1889). The latter disease is caused by the gonococcus, a microbe found by Albert Neisser (1855–1916). The spirochete causing syphilis was identified by Fritz Schaudinn (1871–1906) and Erich Hoffmann (1868–1959) in 1905. Scientific work has since been hampered by the notorious difficulty of growing the spirochetes in culture. Investigators recently determined that this is due to the fact that the bacterium has very few sets of enzymes for building complex molecules essential to its life—proteins, complex lipids, and even nucleotides—and must "steal" them from the host.[6]

Until the beginning of the twentieth century, only two agents were used in the treatment of syphilis: mercury and potassium iodide. Mercury was used in a number of ways: in topical preparations, as vapor to be inhaled, or in mixtures for oral administration. Calomel, or mercurous chloride (Hg_2Cl_2), entered into the composition of many antiluetic medicaments. However, mercury is highly toxic, and its antibacterial potency is limited. The sweating, salivation, and diarrhea that it produces are manifestations of toxicity but were wrongly interpreted as evidence of its efficacy. Potassium iodide was extolled by the great clinician William Osler (1849–1919), who wrote that "under its use ulcers rapidly heal, gummatous tumors [lesions characteristic of syphilis] melt away...an action only equaled by that of...iron in certain forms of anemia, and by quinine in malaria."[7]

The influence of syphilis in deciding the outcome of historical battles is undeniable. Generals have long known that venereal disease is a fearsome enemy, under whose thrusts whole battalions may be undone. It saps the fighting men's energy, enfeebles the troops, and undermines the army's morale. Yet soldiers away from home, stressed by the misery and despair of war, are naturally drawn to the escape that the intense physical pleasure of sexual encounters can afford them. Nor are these encounters rare, when many women in the areas of conflict, pressured by dire scarcities and atrocious suffering, view prostitution as an opportunity to gain financial security. Medical officers in the two world wars commented with dismay on the "disgraceful" spectacle of swarms of women waiting for soldiers on leave, who themselves could hardly wait to spend their pay in the most predictable ways.

In the First World War, the commanders of the Allied forces had reason to fear that the ravages of venereal disease (commonly abbreviated as VD) might become decisive in the campaign. To control its spread, they used a mixture of pragmatism and bigoted morality. The Canadians fared worst, but condoms were not distributed to the men, on the premise that soldiers thus provided would feel incited to engage in sexual intercourse, because pregnancy could also thus be avoided. Instead, they were given a "preventive kit" containing potassium permanganate and a tube of 30 percent calomel ointment, to be applied after exposure. British commanders, actuated by no less strict principles, shunned the measure, on the grounds that a man with a packet was likelier than one without to fall into sinful behavior. Both British and Canadian commanding officers preferred such policies as programs of sports and physical activities designed to keep a man's mind away from lustful thoughts; frequent and inspiring hortatory speeches to remain chaste; the suspension of disability pension for those infected, on the grounds that the disability had been "self-inflicted"; and even pay deductions for the time they were demobilized and under treatment![8] There is little evidence that any of these measures worked.

The German chemist Paul Ehrlich (1854–1915) worked with arsenic compounds, which had been known to have some antispirochete potency. After literally hundreds of trials, on attempt number 606, Ehrlich and his assistant, the brilliant Japanese bacteriologist Sahachiro Hata (1873–1938), synthesized a derivative, patented under the name of Salvarsan 606, with undeniable therapeutic value. Although this was a remarkable advance, the new drug, even after refinements and modifications, was toxic to the liver, its administration was painful, and it was ineffective in curing neurosyphilis, the crippling, most dreadful complication of lues. The world had to wait for the identification of penicillin in 1929 by Alexander Fleming (1881–1955)—a discovery arrived at serendipitously, like so many others in science—before seeing an antiluetic medicament that was both powerful and nontoxic.

Much has been written about the events that led to the discovery of penicillin, which was included by the medical historians Meyer Friedman and Gerald Friedland as one of "medicine's ten greatest discoveries."[9] It all started when Fleming noted that the bacteria he maintained in a culture dish failed to grow in the vicinity of a fungal colony that had (accidentally) contaminated the culture. Fleming's laboratory happened to be in close proximity to another one where fungi were cultured. In a stroke of good luck, the door of his lab had been left open (though it was usually closed), creating the draft that conveyed the fungus. Fleming decided to go on vacation, leaving the Petri dish outside the incubator, which was singularly fortunate, for inside the incubator the bacteria would have proliferated vigorously, thus inhibiting the fungal growth. That the few fungus spores that contaminated the cultures happened to belong to a penicillin-producing species seemed providential, since not all members of the genus *Penicillium* have this ability. And that the bacteria in the cultures were sensitive to penicillin was no less wondrous an occurrence, considering that many strains are naturally resistant to this antibiotic.

In sum, it was a concatenation of unbelievably lucky circumstances—the right spores, falling at the right time, in the correct amount, on the right bacteria—that brought about the growth inhibition. Of course, the presence of an observer able to perceive and interpret the phenomenon was not the least of the fortuitous happenings. Still, this was not enough. Before penicillin could begin saving millions of lives, it had to be refined, purified, tried on different bacteria, assayed for potency and toxicity on animals, and mass-produced. This came about through the efforts of many people, including the Australian-born Howard Walter Florey (1898–1968) and the German-born Ernst Boris Chain (1906–1979).

Both shared with Fleming the 1945 Nobel Prize in Physiology or Medicine.

"War is good to medicine"—alas, this cynical dictum contains more than a kernel of truth in it. Had it not been for the war, the combined scientific and industrial might of Britain and the United States would not have coalesced to bring about the mass production of the first antibiotic in so short a time. All workers were under great pressure during World War II, as Germany was possibly on the same track. At one point, Florey, Chain, and other scientists rubbed the penicillin fungus on the inside of their clothing, as a precautionary measure: in case of a German invasion, they would be able to escape the Nazi grip carrying the precious fungus with them and continue their work in a safe haven. Today, a new drug could not possibly emerge so quickly. Under present regulatory constraints, the testing and assaying of a new drug are extremely involved and take many years.

As it was, there was now a cure for pneumonias, osteomyelitis, deadly infections of the cardiac valves, streptococcal throat infections—a host of diseases for which there had formerly been no treatment (all physicians could do was hope the patient's own immune system would overcome the invading germs). Ironically, in the early trials of penicillin, when its ability to combat infections was established, no one assayed it against syphilis. Otherwise, they would have seen its overwhelming potency against the spirochete.

The number of syphilis cases has been brought to an all-time low. But are we truly better off today? The U.S. Centers for Disease Control and Prevention estimates that in this country there are 15 million *new* cases of sexually transmitted diseases each year.[10] Today there is no single epidemic but no fewer than twenty-five diseases that are spread chiefly through sexual activity. Not including AIDS, the three most common diseases are human papillomavirus (the cause of venereal warts), with 5.5 million new cases every year (a total of 20 million individuals are thought to be infected); chlamydia (3 million new cases each year); and herpes (1 million new cases annually; 45 million people are estimated to be infected). The magnitude of the problem clearly shows that adequate control cannot depend solely on the development of a pharmaceutical "magic bullet" but is also contingent on better education—the CDC estimates that one quarter of all cases affect teenagers—and the creation of social conditions that foster safer lifestyles and decisions.

SMALLPOX

Smallpox (variola) is a disease characterized by fever, followed a couple of days later by a generalized skin rash that is at first papular (raised lesions), then vesicular (fluid-containing blisters), and finally pustular. The skin lesions heal, leaving pitted scars that often disfigure the face and that may affect the eyes, producing loss of vision. Although minor forms of milder course exist, smallpox is a serious disease, with a 20 to 40 percent mortality rate. This malady has been known since antiquity. In the Cairo Museum, the mummified body of Pharaoh Ramses V (died 1156 B.C.) exhibits lesions that appear to be characteristic of smallpox. An ancient Sanskrit text of India describes the disease, and in Hinduism a goddess of smallpox, Shitala Mata, has been worshiped for centuries. In China there is documentary evidence that smallpox has existed since at least 1122 B.C. In some cases the rash is preceded by a blush, similar to what is seen in measles. Rhazes (c. 865–c. 925 A.D.), the great Persian physician, lucidly distinguished between these two diseases in one of his minor treatises.[11]

Smallpox, in its most deadly form, came to America in 1518 with the Spanish conquistadors. First, the Mayan communities were nearly annihilated; next, the Aztec society was smashed. Europeans, by virtue of their living conditions, had been subjected to greater viral exposure than had the Mesoamerican Indians, whose lack of immunity caused unimaginable destruction in the native populations. The disease spread rapidly to South America, where the Incas were decimated, afflicted with excruciating bone pain, hematuria, and hemorrhages in the eyes—all symptoms not usually present in Europeans. The Aztec name for smallpox was *hueyzahuatl,* meaning "the great leprosy," as the victim's body was completely covered with pustules.

The native Mexicans had used a variety of remedies for skin rashes, which they tried to use on smallpox. They rubbed bitumen on the diseased part; they blew tobacco smoke or poured infusions of various plants on it; they consumed hallucinogenic drugs; they even tried pouring *pulque,* the fermented drink from the agave plant that is consumed to this day in Mexico, on the skin lesions. Nothing worked. They also had to contend with other infectious diseases, such as measles, mumps, and tuberculosis—"the diseases of civilization"—to which they had little or no immunity. Some estimate the population of the whole Aztec Empire, when the Spaniards arrived, at about 30 million. Half a century later, in 1568, there were only 3 million. In the Valley of Mexico alone, the pre-Columbian population had been 1 million to 2 million. By the year 1650, after more than one century of Spanish colonization, only 70,000 Indians remained. Many died from the effects of enslavement, malnutrition, and ill-treatment, but the chief cause of death was infectious disease, especially smallpox.

Even if these mortality figures are somewhat exaggerated, the total loss of life still surpasses the estimated number of deaths during the genocidal persecutions in Europe during World War II. Thus, the conquest of America was owed not so much to the often-rhapsodized valor of the conquistadors as to the deadly pathogenic viruses with which they waged bacteriological warfare.

Immunization to smallpox is possible by giving a susceptible person material taken from smallpox lesions, a practice that is called variolation, and has been known for a long time. In China, during the Song Dynasty (960–1280), scabs from healing skin lesions were ground into a dust and blown into the nose through a silver tube. In various parts of the world, mothers were known to feed material from skin lesions to their children, in an effort to preserve them from the disease. Arab physicians practiced variolation by rubbing the fluid from a smallpox blister into a previously made scratch on the patient's arm. The problem with these methods was that the subject sometimes acquired the full-blown disease and could die from it. After variolation the mortality rate was cut in half—undoubtedly an improvement, but still too high a figure.

Reports of the practice of variolation had reached the Western world, but the procedure did not become generalized until the eighteenth century. Instrumental in its adoption was a remarkable woman, Lady Mary Wortley Montagu (1689–1762), daughter of the fifth earl (later first duke) of Kingston, marquess of Dorchester.[12] She was intellectually gifted, a prolific writer, and beautiful. Her resplendent grace and charm were captured in an oil painting by Sir Godfrey Kneller (1646–1723), the leading portraitist of the crowned heads of Europe at the time. This lady, of whom the poet Alexander Pope (1688–1744) wrote that in her person "every grace / with every Virtue's joined," was disfigured by smallpox at twenty-nine years of age. She was left with her face scarred and devoid of eyelashes. This gave her a harsh appearance, most unlike her placid air, which had once been celebrated by the poets. Thus changed, she could no longer keep her former admirers in a state of adoration. Pope, who had apparently made advances that had been shunned, now compared her to Sapho in *The Dunciad,* and Horace Walpole (1717–1797) described her as dissolute and profligate.

Lady Mary became the wife of the British ambassador to Turkey, and while in Constantinople she became convinced of the benefits of variolation. Unbeknownst to her husband, and against the advice of the embassy's chaplain, who believed that inoculation was un-Christian and effective only for heathens, she had her children variolated. Back in England, she persuaded other members of the aristocracy to adopt the same measure, which aroused considerable interest. Eventually, the children of the royal family were inoculated, although not without first trying the procedure on six condemned criminals, who were paid with their freedom in case they survived. All six survived after suffering only a mild form of smallpox. Yet the danger of variolation was undeniable: the two-year-old child of the earl of Sunderland died of complications, as did many other, less privileged persons. This caused serious reservations about the procedure.

In the United States, fear of the complications led to severe restrictive laws in several of the thirteen colonies. Historians have suggested that the lack of variolation among American soldiers led to an outbreak of smallpox during attempts to take Quebec and thus the failure to cut the British supply line in the war of 1812. The English, who had been inoculated, were healthy and repulsed the attack. So it is that smallpox may have played an important role in maintaining Canada within the British Empire.

Variolation was replaced by the much safer vaccination thanks to Edward Jenner (1749–1823), who had been variolated when he was only eight years old. He recalled that this measure, which was supposedly protective, had nearly killed him. This was because the standard medical practice was to "prepare" the subject with repeated bleedings, purges, and fasting. Jenner recovered and, according to custom, at the tender age of thirteen years was placed as apprentice to an apothecary. Following a seven-year apprenticeship, he studied with the renowned surgeon John Hunter for two years. Under the guidance of his illustrious mentor, Jenner would demonstrate great scientific talent and uncommon productivity. A naturalist of wide-ranging interests, he made contributions in areas as diverse as the natural history of some birds of the region and the role of human arteriosclerosis in angina pectoris.

But it was his insight into vaccination and the prevention of smallpox that would bring him everlasting fame.[13]

It was during his apprenticeship with the great surgeon that he first heard from country folk that milkmaids who got cowpox, a disease of the udders and teats of cows, became resistant to smallpox. Milkmaids contracted cowpox by touching the cows; they developed ulcers of the skin of the hands, but these lesions were not accompanied by severe systemic symptoms and did not seem to be life-jeopardizing. Later, in his years as a physician practicing in the English countryside, he had plenty of opportunities to confirm that cowpox conferred immunity to smallpox. Moreover, he saw that variolation caused little or no reaction in people who had been infected with cowpox; and, of even greater interest, that person-to-person transmission of cowpox also protected against smallpox. From this observation he derived the idea of administering cowpox as a protective measure. Jenner's "cowpox inoculation" was later named, appropriately, "vaccination" (from the Latin *vacca,* the word for "cow," from which comes *vaccinia,* "cowpox"). The precise date on which Jenner performed the crucial experiment was May 14, 1796. On that day, he gave the first vaccination to an eight-year-old boy, James Phipps, son of a laborer who worked on the Jenner estate, by scratching his arm with a lancet previously dipped in the fluid from a cowpox lesion of a dairy milkmaid.

The result was entirely as predicted. The vaccinated child was now immune to smallpox. Jenner, not content with this single observation, inoculated other children with fluid obtained from the harmless pustule that develops at the site of vaccination (later, in a book devoted to this topic, he stated that the ideal time to collect the fluid was between the fifth and eighth days). He was so sure of the procedure that he included his own son, Robert, among the children he vaccinated.

The widespread adoption of vaccination did not come about as swiftly and smoothly as such a momentous discovery seemed to require. Envy, ignorance, and bad faith incited preposterous claims that animal traits or animal diseases appeared in vaccinated persons. Inaccurate reports that the inoculation was ineffective or dangerous held back the widespread application of the vaccine. But neither satires nor envious obstacles impeded the eventual worldwide realization that there was, at last, a way to actually *prevent* a disease. Today we know that a person may be rendered immune by receiving specific protein molecules (immunoglobulins) that attack or neutralize invading

Edward Jenner administers one of the first vaccinations to a child, in 1796.

germs, i.e., antibodies. But these antibodies are effective only for the time they remain intact inside the body, usually three to six months; they degrade as they circulate. In contrast to passively received protection, vaccination works by actively promoting the formation of antibodies by specialized cells or the proliferation of cells whose function is to seek out and destroy invading microorganisms.

Pasteur, Koch, and the Fight Against Tuberculosis

The second half of the nineteenth century is considered the golden age of bacteriology. Medical progress had been irregular, zigzagging, a "patchwork" of empirical notions and scientific theories. The emergence of medical bacteriology, championed by Louis Pasteur (1822–1895), Robert Koch (1843–1910), and a host of other remarkable investigators, would change that.

The life and work of Pasteur have been the subjects of an over-whelming amount of literature.[14] What is most amazing about this famous scientist is the large number of disciplines in which he excelled. From ages twenty-five to forty (1847–1862), he studied chemistry, optics, and crystallography. His work with the molecular constitution of crystals and its effect on polarized light led him to analyze the crystals of tartaric and paratartaric or "racemic" acid, products encountered in the sediment of wine during fermentation.[15] In the course of this work, he realized that living microorganisms—yeast cells or bacteria—are indispensable to fermentation. Accordingly, he next focused on the culturing of bacteria and the specificity of their action. He became, in effect, a biological chemist.

To appreciate the revolutionary nature of this work, one must remember that the prevailing opinion of the scientific community at the time considered fermentation, like the putrefaction of organic matter, to be the consequence of purely chemical reactions. Scientists knew that yeasts or bacteria could be observed in sour milk, in spoiled beer, or in wine during fermentation. But these microorganisms were interpreted as the consequence, not the cause, of the respective processes. To propose that living organisms were responsible for the observed changes sounded like vitalism. It was like bringing in a mysterious life force to explain what all the leading authorities interpreted in strictly physico-chemical terms. Thus, the work of Pasteur seemed reactionary—an attempt to set science back fifty years—and he was met with ardent opposition from the most respected sectors of the scientific establishment.

Of course, there was also the question of where the bacteria that caused fermentation came from. Here, the whole question of spontaneous generation resurfaced. This error had not been completely expunged, despite incontrovertible demonstrations of leading scientists in the preceding two centuries. Pasteur's rigorously controlled, precise experiments put the matter definitively to rest in 1860. He was not timid in confronting his opponents. His lucid expression, charged with the force of unassailable truth, flashed forth in debates; the records of contemporary discussions in learned societies show him wielding this eloquence with crushing effect on his adversaries—a sure way to create envious enemies.

From 1862 to 1877 he was primarily a biologist. Having demonstrated the preeminent role of bacteria in fermentations, he developed a means of eliminating them by heating the liquid to 55°C. This became known as "pasteurization," now universally applied in the preservation and transport of milk and other foodstuffs. In 1865, the silk industry of France and other countries suffered great losses due to a disease that affected silkworms, called *pébrine* (from a Provençal word, *pébre,* "pepper," because the silkworms developed peppered stains on their bodies). Pasteur, now turned entomologist, showed that it was due to microorganisms and that it was contagious. He also identified another disease, called *flacherie,* and, although unable to isolate the respective causal microorganisms, he taught silk workers how to identify the affected worms and how to avoid the spread of these diseases, thereby saving the silk industry from total collapse. This discovery saved the French government more money than the sum owed to Prussia as reparations from the lost war against that nation.

Having demonstrated that fermentation and putrefaction are due to living microorganisms, Pasteur was questioned about whether the festering or suppuration of wounds and ulcers stemmed from a similar cause. Physicians opposed this view; "the diseases that the ancients called putrid" (an expression used by Pasteur in one of his speeches) could be only "a work of death," that is, explainable as a chemical decomposition of dead tissues. And this man had the impertinence to state that they were "a work of life"! So Pasteur once more had to use his powers of persuasion. He had discoursed before the Academy of Sciences, and now he appeared before the Academy of Medicine. The strength of his demonstrations once more carried the day.

Not the least of Pasteur's astounding talents was his knack for public relations. Critics have seen this as unbecoming showmanship. He endeavored to produce vaccines (Pasteur insisted in so naming the preparations meant to prevent bacterial diseases in honor of his predecessor, Edward Jenner) against chicken cholera and anthrax. This he did with scarcely any mention of Robert Koch, the man who had painstakingly and brilliantly shown that anthrax is caused by a bacillus that infects animals as well as human beings.

Pasteur devised ways of making vaccines with "attenuated" bacteria, as he called germs whose capacity to induce disease was lessened through aging, growth in deficient media, or some other means but that still retained their capacity to induce immunity. His work had yielded what seemed a promising vaccine against anthrax, although its effectiveness had yet to be extensively tested. Then, smarting from the skepticism of critics who doubted the validity of his findings, he agreed to organize a public demonstration. In Pouilly-le-Fort, the rural community in which the event took place, sheep and other animals were injected with Pasteur's vaccine against anthrax on May 5, 1881; an equal number were left uninjected, as controls. The vaccine was "boosted" by second and third injections weeks later. Then, with characteristic showmanship, in the presence of journalists, local politicians, physicians, sheep raisers, veterinarians, and sundry curious witnesses, a lethal dose of a virulent strain of anthrax bacteria was given to all the animals. Forty-eight hours later, all the inoculated animals, except those that had been vaccinated, lay dead or dying. The results could not have been more stunning.

Pasteur's savvy in matters of publicity and self-promotion paid off grandly. The dispute between Pasteur and his critics had been highly publicized from the beginning. Journalists from America observed the proceedings; the correspondent for *The Times* of London sent daily dispatches to his newspaper. The stakes were inordinately high. The vaccine was still not quite perfected. Pasteur had risked the possibility of massive public ridicule. A number of things could have gone wrong. Had he chosen the wrong animals or had the injection dosage not been properly calculated, the results would have been inconclusive, his critics emboldened, and himself grievously ridiculed. As it was, he had scored a resounding triumph.

Pasteur was already the most revered scientist in France; now he became an international icon. People flocked from all over the world to ask for his advice on the containment of animal or plant epidemics and, as happens to the famous, even on matters unrelated to his expertise. But the most spectacular of his discoveries had yet to come. A few years later, he turned his attention to a disease much dreaded since antiquity for its horrifying symptoms and usually fatal end: rabies. Pasteur suspected that this disease was caused by a microorganism, but his efforts to detect it consistently failed. No wonder, since the causative virus could not be seen with the best optical microscopes available in his time nor even those made today: none can provide the necessary magnification; seeing the virus was not possible until the invention of the electron microscope. However, with the able help of Pierre-Émile Roux (1853–1933) and Charles-Édouard Chamberland (1851–1908), he induced rabies in rabbits and realized that the spinal cords of these animals were quite virulent when fresh but seemed to lose their infectivity when left to dry for over fourteen days. If dogs were injected first with the "attenuated" material and then given inoculations of gradually increasing virulence, they became resistant, i.e., immunized, against rabies.

There were still important facts to ascertain. It was not known, for instance, whether the injected virus could have serious side effects. The appropriate dose for human beings had yet to be worked out. Even the circumstances under which to give the vaccine were not known, since not every person exposed to the virus became sick. Uncertainty surrounded all these important questions, when, on July 6, 1885, a nine-year-old boy named Joseph Meister was brought to Pasteur after having suffered fifteen bites from a rabid dog. Meister's doctor was aware of Pasteur's research, and, seeing the ghastly wounds that the boy had sustained and fearing the worst, he recommended that the child be taken to Pasteur. Pasteur took yet another risk. He administered a series of painful injections of gradually increasing virulence, just as he had done with his experimental animals. The desperate measure worked: the boy remained well. Shortly thereafter, a second patient was treated similarly, with equal success.

Rabies, which until then had been untreatable, had now been effectively prevented for the first time. Pasteur, ever the showman, exploited the publicity, magnifying the benefits of his research and making it appear much more solidly grounded and definitive than it actually was. But whatever human foibles may have attended his personality, Louis Pasteur, a nonphysician, looms large in the history of medicine. His magnificent contributions were unequaled in his time, and arguably still are in ours, and Pasteur died amid universal acclaim. He was buried in the prestigious Pasteur Institute, which he founded in 1888. Joseph Meister, the boy he had saved from certain death, became the gatekeeper and remained in this post for the rest of his life.

Whereas Pasteur was colorful, flamboyant, and somewhat eccentric, Robert Koch was sober, staid, and, according to some biographers, just plain dull. But the unprepossessing German physician was a born researcher, endowed with a perfectly well ordered mind and a fixity of purpose. Working alone, in a small town, far from the great scientific and cultural centers, availing himself of only a microscope (a gift from his wife) and a few tools of his own making, he divided his time between his medical practice and his backroom research, building by himself the whole foundation of medical bacteriology. After years of patient labor, he discovered how to grow bacteria in liquid and solid media. Using this methodology, he established that anthrax in sheep (and in humans) was caused by a bacillus; that this bacterium could change into resistant forms, called "spores," able to withstand inclement weather; and that spores lodging in the grass, when ingested by pasturing sheep, communicated the disease to them. At the time, the very idea that human diseases were caused by microscopic living organisms seemed unbelievable, making Koch's discoveries all the more significant.

In order to prove that a bacterium was the cause of the disease, Koch not only had to locate the suspect microorganism in the patient's body, he had to recover it from every similarly affected subject. Furthermore, the microbe had to be grown in pure form. But this was not enough. He demanded that the bacterium, when injected into a susceptible animal, be capable of reproducing the disease. Only when these exacting requirements were fulfilled—they came to be known as "Koch's postulates"—was he willing to concede a cause-and-effect relationship between the microorganism and the pathologic state.

Koch's greatest glory was achieved in the struggle against tuberculosis, also known as "phthisis" (from the Greek *phthinein*, "to decay, to wane"), "the chest disease," or "consumption" during the industrialized nineteenth century. In France alone, an estimated 9 million deaths were attributable to tuberculosis in that century. The victims tended to be young, as portrayed in a number of novels, plays, and romantic narratives of the period. Many famous artists died from tuberculosis, Frédéric Chopin and members of the Brontë family among them. Among the doctors, Marie-François-Xavier Bichat (1771–1802) and René-Théophile-Hyacinthe Laënnec (1781–1826) succumbed to "the consumption." The young poet John Keats (1795–1821) had some medical training and recognized the seriousness of an episode of hemoptysis (coughing up blood) that he suffered. He poignantly exclaimed, "I know the color of that blood. It is arterial blood. I cannot be deceived in that color. That drop of blood is my death warrant. I must die." Among the poor, the devastation was extreme, but the disease spared no age, gender, or social class.

The disease manifested itself in cough, fever, sweats, progressive debilitation, and eventually the spitting of blood due to extensive destruction of lung tissue. The clinical symptomatology had already been minutiously described. Auscultation of the chest was raised to the level of a fine art, especially after Laënnec's invention of the stethoscope in 1816. Chest sounds, normal and abnormal, became matters of serious medical study. Yet the origin of tuberculosis vexed all investigators. Curiously, Laënnec, who died of this disease, believed it to be a hereditary condition. Koch proved that the cause was a bacillus, *Mycobacterium tuberculosis*. He devised a method of obtaining the bacterium in pure culture by adding blood serum to the culture medium. Anyone familiar with bacteriological work knows how difficult it is to grow this microorganism and can appreciate the ingenuity needed to overcome such an obstacle. Faithful to the rigorous standards he had set himself, he reproduced the disease in experimental animals. His demonstration was unassailable and corroborated one of medicine's greatest discoveries. He reoriented medical thinking in respect to that most deadly malady and provided a rational basis for prevention.

Knowledge of the origin of the disease led to the development of a vaccine made from attenuated tubercle bacilli from cows, known as BCG (abbreviation for "bacillus Calmette-Guérin"), named after Albert Calmette (1863–1933) and Jean-Marie Guérin (1872–1961) of the Pasteur Institute. But perhaps of even greater benefit was the systematic application of Koch's methods and rigorous scientific principles by his followers and disciples. Thanks to this approach, the bacteria responsible for many diseases, such as typhoid, pneumonia, gonorrhea, plague, undulant fever, tetanus, syphilis, and whooping cough, as well as streptococcal and staphylococcal infections, were identified in rapid succession.

Showered with honors and official distinctions, Koch traveled widely, charged with studying various epidemics. Then, first in an 1890 speech at the Berlin International Congress of Medicine and afterward in writing,[16] he made the sensational announcement that he had found a cure for tuberculosis in a substance he called "tuberculin." It was a most lamentable faux pas. He exhibited the very failings he had reproved in Pasteur: the rush to proclaim definitive results in work still in progress; the fondness for publicity and self-promotion; the desire to impose his authority over any opinion contrary to his own.

Parenthetically, it must be said that medicine was suffused with the strong nationalism of the times, and the pain of the Franco-Prussian

War (1870–71) was fresh in the memory of the two former foes. Pasteur did not hide his antipathy for the Germans, and Koch, in many ways the prototypical Prussian, reciprocated the sentiment with a profound disdain for the Frenchman. Pasteur worked on anthrax, hardly acknowledging his great debt to Koch; the latter, in turn, lost no chance to disparage or discredit his rival's findings.

"Tuberculin" was useless as a remedy. It was nothing but a filtrate of *Mycobacterium tuberculosis* cultures. Thousands of patients received the supposed medicament, but their hopes were dashed, as no one was cured. It was rumored that Koch had sold the patent to a pharmaceutical company for millions of marks, which he used to pay the expenses incurred in his divorce and his remarriage to a young lady many years his junior.

Tuberculin eventually found a medical use: not as a cure but as a diagnostic skin test. This was discovered by Clemens von Pirquet (1874–1929), an early student of immunology. Individuals who have been previously exposed to tuberculosis, when injected with tuberculin, display several days later a visible inflammatory reaction (an example of a complex biological phenomenon known as "delayed hypersensitivity"). In the nineteenth century, this reaction was widely used to eliminate cattle infected with tuberculosis. This was very important, because unpasteurized milk containing the bovine strain of the bacterium can cause infection, entering the human body through the tonsils.

In the developed world, mortality due to tuberculosis was reduced by 90 percent. Effective drugs now exist, and many in the medical profession would like to claim credit for this improvement. But there is evidence that the incidence of tuberculosis started steadily declining long before effective antituberculous medication became available[17] (streptomycin was first produced in the 1940s). It may be that the benefit was largely due to improvements in living conditions: greater wealth, better nutrition, isolation of infected patients, superior hygiene—these factors, rather than a specific medical intervention, may have accounted for the observed effect. The same may be true of other infectious diseases, as the distribution of piped, bacteria-free water, the collection of refuse, closed sewers, and less crowded dwellings no doubt played a large role in containing the spread of contagious diseases.

INFLUENZA

Every few decades, influenza wreaks havoc among human beings. The most intense of the outbreaks surged during World War I, in 1918, and blazed for a couple of years. Half the world population became infected, and, according to recent estimates, 50 million to 100 million people died.[18] Given these numbers, influenza qualifies as the most lethal of known epidemics, surpassing bubonic plague, syphilis, and AIDS.

Although it was popularly known during World War I as the "Spanish flu," this is a misnomer. Spain was not a contender in the war, so its press could devote more space to the epidemic than countries engaged in the conflict could. Therefore, the epidemic became more strongly associated with Spain in the public's mind than with any other country. When the disease reached Russia, Moscow journalists jokingly wrote a headline saying "The Spanish Lady Has Arrived!" and the name stuck. In reality, the first case was traced to Camp Funston, Kansas, in March 1918.

Affected individuals complained of fever, chills, headache, prostration, dry cough, and worsening difficulties of respiration that could produce cyanosis (bluish discoloration of the lips, nail beds, and other areas, due to inadequate blood oxygenation). Death came swiftly, often from secondary bacterial pneumonia. No biological reason has been found that can explain the extraordinary severity of the 1918–20 pandemic. This important question has intrigued researchers, to the point that several expeditions have been organized to retrieve tissue samples from persons known to have died of this disease in an Arctic encampment. The human remains of victims of the "Spanish flu" buried in permafrost were thought to be sufficiently well preserved to permit study with modern techniques of molecular biology. Thus far, however, it has proven impossible to ascertain the reason for the high lethality of the 1918–20 pandemic.[19]

The need to continue research in this area should be obvious. Periodic scares emphasize the urgency of this concern. In 1997, there was a limited outbreak in Hong Kong, in which the virus H5N1 was transmitted from birds to humans and a third of the people infected died. Two million chickens had to be destroyed. There was a new outbreak in Asia in 2005, also caused by influenza virus H5N1. Because human beings have developed no immunity to the virus, the concern was great that a new pandemic might develop. There has been no great human mortality after the migratory birds carried the virus to three continents, but experts warn that, sooner or later, if historical experience and the facts of biology are any indication, a new influenza outbreak will arrive. The question is only when and where it will occur. It is important that we be prepared, but at present our readiness is insufficient.[20]

MISCELLANEOUS CONDITIONS AND FINAL CONSIDERATIONS

Since Jenner's discovery, advances in technology have produced various kinds of vaccines. Some consist of genetically altered or otherwise modified ("attenuated") bacteria or viruses that can still live and reproduce inside a person and therefore stimulate immunity but do not produce disease. Vaccines of this kind include those for smallpox, yellow fever, polio (the oral vaccine), typhoid fever, measles, mumps, and rubella. Other vaccines are made of dead microorganisms, thus incapable of reproducing in the host. These include whooping cough, cholera, meningococcus, pneumococcus, poliovirus, influenza, and hepatitis A and B. Vaccines can also be produced from toxic products derived from bacteria or viruses and suitably modified ("toxoid"). The classic examples are those against diphtheria and tetanus, for in these diseases the harmful effects are due not as much to the bacteria themselves as to the toxins they produce.

Scientific and technological progress has increased our life expectancy, but it has not made us invulnerable. Infectious diseases remain the primary cause of death worldwide. The remarkable advances in bacteriology and the production of vaccines generated some complacency in the 1960s and 1970s. This overoptimism was shattered when physicians in the United States first recognized, in 1981, the existence of acquired immunodeficiency syndrome (AIDS). This disease is characterized by severe breakdown of the immune system, so that patients die of opportunistic infections, often more than one at the same time. Unusual conditions in otherwise healthy individuals, such as Kaposi's sarcoma or *Pneumocystis carinii* pneumonia (PCP), called attention to these patients. Isolated cases had been seen before, at least since the late 1950s and perhaps earlier. Soon the cause was shown to be the human immunodeficiency virus (HIV), a member of the retrovirus family. Retroviruses have been known since 1911, when Peyton Rous (1879–1970), an American pathologist, discovered that one of these agents, when injected into chickens, induced malignant tumors (Rous sarcoma virus). However, the production of an AIDS vaccine was hampered by the swift adaptive mutation of the virus, similar to the high mutation rate of the influenza virus.

As in previous epidemics of history, the spread of AIDS elicited irrational fears and behavior. Because at first the disease was most prevalent in homosexual men and in drug addicts who shared the use of needles, moralizing and victim blaming long antedated social efforts to stave off the outbreak. Some claimed that AIDS was divine punishment for men's immorality (a recurrent idea with every plague); others, that the virus had been created by scientists in secret laboratories of Russia or the United States as a weapon of bacteriological warfare, which had accidentally escaped. Another theory went that the virus had come from outer space. Sadly, one in seven African Americans today believes that the virus was created in a government laboratory in order to control the black population, and one in three is convinced that a cure exists but is being withheld from the poor.[21] These misconceptions are rooted in the racism that blacks have endured in the United States, in particular the infamous Tuskegee episode, in which treatment of poor black sharecroppers with syphilis was purposefully withheld so that researchers could study the evolution of the untreated disease.[22]

More credible was the theory of a British journalist who, after years of painstaking research, published a voluminous book in which he maintained that the AIDS virus was derived from a chimpanzee virus present in tissues used in the production of oral polio vaccines and that this simian virus was unwittingly transmitted to African children, perhaps through a sore or a cut in the mouth, in the early trials of polio immunization.[23] The book caused a considerable stir, because its central thesis implicated the developers of the polio vaccine and those who administered it to the African population. Most scientists believe that the AIDS virus originated in the deep forests of Africa in a simian precursor and that it passed to humans by "natural transition" once social and economic conditions increased contact between apes and human beings (possibly through contamination by blood, for instance while cutting infected chimpanzee meat or through a bite by an ape).

The theory implicating the polio vaccine, based on purely circumstantial evidence, can be neither proven nor disproven but ought to serve as a warning that all precautions should be taken in the development of vaccines.

Plagues are diseases of civilization. History shows that disorders once thought to be newly emerging were, in reality, old, surprisingly predictable, and in large part the result of human activities. Today's situation is no different. Scientists agree that urbanization, deforestation, and contamination of lakes, rivers, and the atmosphere are some of the man-made factors that cause global warming. These factors also promote epidemics, because vectors and their associated diseases, previously confined to certain regions of the planet, are now extending their geographic distribution. In doing so, they encounter hosts not previously immunized. This is a perfect setup for epidemics. Some examples are already on record. In Honduras, indiscriminate deforestation has brought about an increase of the median ambient temperature, which is linked to a 23.5 percent increase in vector-borne diseases such as yellow fever, leishmaniasis, and Chagas' disease. Malaria incidence has risen 425 percent in four years.[24]

Malaria continues to be one of the major killers in the world. Two thirds of the global population is at risk, but 90 percent of deaths from this disease occur in sub-Saharan Africa. A vaccine does not exist, but the full genome of the parasite has become known. This gives hope that new strategies for treatment and prevention will soon be developed, since gene-based vaccines have advantages over those developed using the traditional empirical approach.[25] Malaria is one of the diseases likely to extend its ravages in a warmer world.

Infectious pathogens that for centuries have been in a stable host-parasite relationship with animal reservoirs emerge to attack man when their environmental niches are disrupted. Lyme disease, for instance, had long existed in North America, but the number of cases increased nineteenfold as more and more people came into contact with the vectors (ticks) that transmit the disease. Similarly, colonization of the Amazon basin brought humans into contact with arenaviruses that theretofore had existed only in the depths of the rain forest, causing outbreaks of "new" hemorrhagic fevers. Some diseases may be due to new importation of old microbes, as happened with outbreaks of cholera in South America.

On the other hand, bacteria not formerly known to cause disease have also been identified as pathogenic, as happened with *Legionella*. Before 1976, bacteria of the genus *Legionella* were responsible for sporadic outbreaks whose cause was not determined due to the difficulty in growing the microorganisms and reproducing the disease in laboratory mice. Then, on July 21, 1976, at a convention of the American Legion in the Bellevue Stratford Hotel in Philadelphia, Pennsylvania, 221 conventioneers and other people present were affected by a mysterious disease that manifested with fever, dry cough, and pneumonia; one month later, 34 of these patients had died. The seriousness of this episode prompted a large number of researchers from the Centers for Disease Control and Prevention to join forces and apply the most advanced biomedical technology to discover the cause of the disease. It was found to be bacteria, which resided in the water of the hotel's cooling towers and were spread by the air-conditioning system. These bacteria were thermophilic (they grow at high temperatures), resembling those that live in hot springs in many sites of the world. The knowledge thus acquired allowed the researchers to ascertain that some previous epidemics of unknown origin had, in fact, been caused by *Legionella* species.

In recent history, even diseases thought to be of nonbacterial etiology were revealed as microbial in origin: gastroduodenal ulcers, traditionally linked to anxiety and stress, are now known to be caused by *Helicobacter pylori*. The 2005 Nobel Prize in Physiology or Medicine was shared by the Australian physicians Barry J. Marshall (b. 1951) and J. Robin Warren (b. 1937) for their discovery of the role of this microorganism in gastritis and peptic ulcers.

International travel, migrations, wars, and commerce also increase the likelihood of encounters between pathogens and susceptible individuals. The increase in world population, and the crowding that accompanies it, has the same effect. International trade of foodstuffs makes possible the transfer of pathogenic bacteria, not to mention the transportation of rodents and other vectors that arrive as stowaways in aircraft or sea vessels.

The medical profession itself has also contributed to the emergence of "new" and old pathogens. Bacteria become resistant to drugs by exchanging genetic material with other microorganisms or simply by natural selection after persistent exposure. Overprescription and easy availability of over-the-counter antimicrobial drugs in many parts of the world has contributed to an increase in bacterial resistance. Another cause is the high concentration of patients in hospitals. So-called nosocomial infections are often due to drug-resistant microbes. Human beings may act as reservoirs of bacteria, from which the pathogen disseminates to other susceptible individuals. Reservoirs also act as an ecological niche in which the bacteria persist long enough to develop resistance.

In sum, the development and use of antimicrobial drugs and vaccines, together with better hygiene and improved sanitation and nutrition, caused the remarkable decrease in bacterial diseases achieved in the twentieth century. This bred excessive confidence, now shattered by the realization that bacteria are still major causes of worldwide morbidity and mortality. Bacteria display a marvelous capacity to adapt to environmental pressures. Therefore, new strategies must be developed. If our lives are to be free from the plagues that punctuate the history of mankind, research cannot relent. As the distinguished scientist and Nobel laureate Joshua Lederberg put it, "It is our wits against their genes."[26]

6

CONCEPTS
OF DISEASE

Traditionally, diagnosis is defined as the art of identifying a disease from its signs and symptoms and distinguishing it from other diseases. Of course, this presupposes agreement on the meaning of the word "disease." In everyday life we take it to be clear cut. But history shows that the concept of disease is neither sharply defined nor immutable.

Diseases are not entities in themselves but largely constructs of society. For example, consider a person with a pigmented skin lesion that grows rapidly and becomes protruding, then ulcerated. With time, the affected person becomes progressively thinner and the lesion rapidly advances. Perhaps we see him cough, spit blood, and develop new pigmented lesions, some adjacent to the first one, others more distant. All these are undeniable, objective facts. Yet it is a leap to claim that this person has a cancer of the skin or a malignant melanoma. The observed facts must be fit into a pathological system, a conceptual frame of reference. A disease is not a given but an interpretation of reality, a conceptualization.

History shows that diseases were not always conceptualized in the same way, varying by group and geographic location. Mirko Grmek,[1] a distinguished philosopher and historian of science, illustrated this point with the Hippocratic theory of fevers. The famous Greek physician lived and practiced in an area of the world where most of the population suffered from malaria. Thus, he was accustomed to seeing some patients with intermittent fevers that recurred every three or four days (*tertian* or *quartan*) and others in whom fever was continuous. Today we know that such periodicity is related to the life cycles of the various species of *Plasmodium,* the parasite that causes malaria. The regularity with which the febrile episodes succeed each other in these patients must have suggested to Hippocrates the idea that it was important to introduce the notion of number into clinical medicine. Hence the Hippocratic emphasis on the number of days that it takes for recurrence or resolution to occur in various febrile diseases. The theory of fevers, Grmek points out, would have been very different if the theorizer had been a Scandinavian, since in areas of the world where malaria is unknown, a physician was never exposed to patients whose temperatures showed regular oscillations.

Moreover, the meaning of symptoms and signs can also be variable. In some regions of Africa, infantile diarrheal disorders are considered healthy, at least in their incipient or mild form, as a kind of natural cleansing of the intestine. In pre-Columbian America, intestinal worms were considered helpful for digestion. (Interestingly, today some investigators maintain that intestinal worms, by enhancing the immune response of the bowel, may be helpful in the treatment of ulcerative colitis and Crohn's disease; studies show that patients fed the eggs of pigs' worms experience some improvement.[2]) The ancient Egyptians deified persons with congenital anomalies. Their god Ptah is sometimes depicted as an achondroplastic dwarf (Ptah-Pataikos). The domestic god Bes or Basu, protector of the newly born, was also affected by some kind of malformative dwarfism.[3] In pre-Hellenic Greece, the god Hephaestus was represented as a child who limped on both feet, but by the fifth century B.C., he was a dignified, bearded man without any physical deformity. This may reflect an evolution of the aesthetic ideas of classical Greece, which no longer tolerated a deformed being amid the gods of Olympus, and it is also an indication that the collective attitude toward deformities had changed.

What is considered disease in one culture may be considered health in another culture, or at a different time in the same culture. In Western medicine, however, certain theories of disease exerted a profound and long-lasting influence. Chief among these is Hippocratic medicine, which lasted for more than two thousand years. Hippocrates (c. 460– c. 377 B.C.) is traditionally called "the father of medicine," although precious little is known about his life. He taught at the medical school in his native island of Cos, in the southwestern Mediterranean, and was a contemporary of Plato, who mentions him in a couple of his *Dialogues.* About sixty works make up what is called the Hippocratic Collection (Corpus Hippocraticum),[4] although it is generally accepted that many of these were written not by Hippocrates himself but by various other authors, and thus the views expressed are sometimes contradictory.

The "humoral theory," one of the main pillars of Hippocratic medicine, existed long before Hippocrates, but he gave it a distinct formulation and wide diffusion. Probably it originated as a corollary to cosmogonic schemes that proposed that the universe was composed of four basic elements: fire, water, air, and earth. The simplicity of this theory appealed to the Greek mind; the Greeks then applied it to medicine. In an attempt at symmetry, they believed that humans, born of the combination of the four elements, were composed of four hu-mors: "blood" (originating in the heart), "phlegm" (the brain), "bile" (the liver), and "black bile" (the spleen).

The ancient Greeks were great poets, logicians, and philosophers, but they were poor scientists. As one Victorian scholar put it, "[The Greeks] used the *Logos* where they should have used the scientific method" and they "tried to explain Nature while shutting their eyes."[5] The tendency to eschew experimentation for the sake of a strictly de-ductive method (which was not always guided by the most rigorous logic) led to some brilliant answers, but also to some errors. They fit everything into the theory of the "four humors," just as there were four cardinal points, four temperaments, four ages of man, four seasons of the year, and so on. In youth, in a sanguine temperament, and in the spring season, blood predominated. In old age, winter, and a "lym-phatic" temperament, phlegm was predominant. In middle age, a bil-ious temperament, and summer, bile dominated. In autumn, as in a melancholic temperament and hot countries, it was black bile (also named melancholy) that made its presence felt above all others.

One was healthy only when the four humors were admixed in per-fect equilibrium. If the proportions changed, disease ("distemper") ap-peared. The character of the malady depended on the superabundance or deficiency of the specific humors. Diagnosis consisted of determin-ing the nature of the disequilibrium; treatment could then be directed at restoring balance. It might also happen that the humors would spill out of their normal reservoirs ("fluxion") and cause distempers. The brain was the site of origin of phlegm, and respiratory troubles were ex-plained as brain-to-chest fluxion of phlegm. Epilepsy, taught Hip-pocrates (in the book *On the Sacred Disease*), was caused by a retention of excessive phlegm in the brain. As proof he cited the large amount of fluid found in the brain of an epileptic goat. This is the sole example of a reference to pathological anatomy in the Hippocratic texts.

Although there were some important philosophical differences be-tween Galenic and Hippocratic medicine, Galen (A.D. 129–c. 199) was influential in continuing the humoral theory. Galen reinforced the idea that the four elements of matter had "qualities," namely, dryness, humidity, coldness, and heat. These qualities existed in various combi-nations: earth was cold and dry, water was cold and wet, air was hot and wet, and fire was hot and dry. Just as disease resulted from imbalance of the Hippocratic humors, it could also be caused by a number of disturbances of the respective qualities. The matter was debated for more than a thousand years before these systems were discarded as useless.

But well into the seventeenth century, at the threshold of the Enlightenment, Hippocratic and Galenic ideas still held sway. Fever was considered not a symptom but a disease in itself. It could be "simple" fever: "a hot distemper that comes from the heart," interpreted as "an effort of Nature to cook the corrupted humors." This disease was combated with "cooling and refreshing foods." A different problem was posed by "hectic fevers," which were more severe. Hectic fevers were treated with humectant foods and prolonged baths, in order to replenish the missing humid principle. Finally there were the sinister "pestilential fevers," dreadful companions of severe infectious diseases such as bubonic plague, typhus, and others. Attributed to an excess of black bile, they were treated with the nearly universal remedy, bleeding.

Bleeding had become a constantly used therapeutic measure. Today, the idea that a debilitated person, including one suffering from the effects of blood loss, should be treated by bloodletting strikes us as the utmost absurdity. But in the context of Hippocratic theory it made perfectly good sense: the body had to be rid of corruptible humors found in the blood. Corruptible matter was attracted to ulcers, wounds, and other suffering bodily sites where there could be stagnation and putrefaction, which resulted in the need for bleeding. It was good for almost everything—not only for treatment but also in the midst of good health, as a preventive measure. There were many benefits of bleeding: "it invites digestion, it evokes the voice, it builds up the sense, it moves the bowels, it enriches sleep, it removes anxiety..."[6] As an old English saying put it, "A bleeding in the spring is physic for a king."

The humoral theory did not go entirely unchallenged. At the beginning of the Renaissance, Paracelsus (1493–1541; his true, Latinized name was Philippus Aureolus Theophrastus Bombast von Hohenheim) attacked it, wishing to replace it with his own ideas. Paracelsus remains a colorful, highly controversial figure in the history of medicine. Aggressive, pugnacious, and loud, he made enemies everywhere he went. At the university of Basel, where he taught for some time, he was a sharp critic of Galenic theories and an unrepentant troublemaker. His lectures were not in Latin, as was the custom for academics in his time, but in the local Swiss German. In his criticism of Galen and Avicenna, the most revered figures of the faculty, he did not balk at using profanities and foul language. He had to leave more than one city in a hurry, under cover of darkness, to avoid being hurt by his enemies.

All but a few excerpts of Paracelsus's writings are difficult to read, due to his frequent references to alchemy and occultism and an utterly sui generis terminology of his own devising. Yet there are valuable insights within this mass of esotericism. First, Paracelsus concluded that diseases have exogenous causes. In contrast to the ancient tradition, which placed the origin of disease within the body (as a disturbance of the humors), he postulated extraneous, disease-producing materials arriving in food and drink. Second, although Paracelsianism is inextricably bound to the alchemy of his time, he showed that disease could be understood as a *chemical* process, and a highly specific one at that. Furthermore, he saw that disease was a local process, preparing physicians to accept the concept of abnormal organs as the primary seat of disease.[7]

Indeed, the humoral theory was not entirely superseded until the anatomic concept of disease was consolidated. But before this could happen, a true revolution had to take place in the prevailing ideas about some of the fundamental facts of human physiology. The most remarkable development was William Harvey's discovery of the general circulation. Harvey (1578–1657) arrived at his monumental discovery through a combination of anatomical knowledge and carefully reasoned observations. He had studied at Padua, as a disciple of Fabricius ab Aquapendente (see chapter 1), who described the structure of the veins in a work entitled "On the little openings of the veins," *De venarum ostiolis* (1603). Thus, Harvey was aware of the existence of venous valves that give way to the passage of blood toward the heart but oppose it in the contrary direction. He had also performed vivisections and seen the large amount of blood that flows through the heart. He concluded that the body could not possibly produce, or metabolically consume, so much blood in a short time. If one assumes—conservatively—that 60 milliliters of blood are expelled from the heart per stroke and the heart beats seventy times per minute, this gives 4,200 milliliters per minute. In a half hour this would amount to 126,000 milliliters, that is, 126 liters. But the body contains only about 5 liters of blood. Harvey's conclusion was unavoidable: the blood had to recirculate. In Harvey's words, this blood volume cannot be furnished by those things which we do take in [i.e., foods and drinks], and in a far greater abundance than is competent for nourishment: It must be of necessity concluded that the blood is driven into a round by a circular motion in creatures, and that it moves perpetually; and hence does arise the action and function of the heart, which by pulsation it performs; and lastly, that the motion and pulsation of the heart is the only cause.[8]

The experiments that corroborated his reasoning may seem surprisingly simple, but in his time it was generally assumed that the blood was constantly produced by the liver, which sent it outward via the veins, not the arteries, and that it was somehow promptly "consumed" in the organs and tissues of the body. The arteries were assumed to pulsate on their own, by an inherent power of contractility, as did the heart, and not necessarily in relation to a wave of blood. The very concept of systole and diastole had yet to be formulated (at the time the filling phase of the heart was thought to coincide with the forward protrusion of the cardiac muscle and the squeezing out of the blood with the heart's wall flattening; today we know it is exactly the reverse: the heart wall bulges anteriorly when it contracts to eject the blood and recedes when it relaxes during diastole). Moreover, there was no agreement as to the volume of blood pumped by the heart with each stroke: a theory of "ebullition" maintained that it was very little, because blood was mainly in the form of froth as it passed through the heart, where it was heated.

Harvey illustrated his experiments in his 1628 landmark book *Anatomical Exercises Concerning the Motion of the Heart and Blood in Living Creatures (Exercitatio anatomica de motu cordis et sanguinis in animalibus).*[9] He observed that a moderately tight tourniquet applied to the arm produces dilation of the veins, as the course of the blood flowing through them is impeded. It also reveals small, localized swellings at the site where venous valves are present. Let the experimenter push with the tip of his right index finger against one of these swellings. Next, while this right finger is still compressing, let the experimenter push with his left index fingertip on the same vein, sliding it upward (toward the subject's shoulder) so as to drive the blood in the direction of the next valve and a little beyond. The venous segment between the two compressing fingertips will appear bloodless, collapsed. Then, if only the compression by the upper finger is removed, the vein will remain collapsed, because blood does not flow backward, being stopped by the valves. But if the lower finger is released, the blood will fill the vein from below. This simple observation clearly showed that blood flow in the extremities goes from the body's periphery to the center via the veins and is aided by the presence of valves. That the pulse disappeared when the tourniquet was tightened further and the extremity became cold and painful if the tightness persisted indicated that the course of the blood flow in the arteries—not the veins—is from the center to the periphery.

Knowledge of cardiocirculatory physiology received an important impetus when Stephen Hales (1677–1761), a scientist who is considered the founder of vegetal physiology, performed the first measurement of venous and arterial pressures in a mare.

William Harvey's landmark researches were truly revolutionary in the history of science. His studies demonstrated that the heart, until then surrounded by mystery and believed to be the seat of life's energy, as well as the site of origin of emotions, was, in fact, an organ amenable to scientific study. Those who endeavored to understand the various functions of the body were emboldened, seeing that an organ until then deemed to possess preternatural qualities was made accessible to reason. Progress in other areas of physiology also helped to shape the modern conception of Western medicine. Among these, the discovery of the digestive function is especially noteworthy.

How foodstuffs are broken down and liquefied in the stomach had long been a disconcerting puzzle. Some hypothesized that the stomach's muscular wall was sufficiently strong to mechanically disrupt and churn the ingesta by beating them and turning them vigorously. Equally fantastic, the stomach was imagined as an oven of sorts, in which raw aliments were softened, cooked, and eventually liquefied.

The systematic work of Réaumur (1683–1757; full name: René-Antoine Ferchault de Réaumur) did much to debunk those myths. A remarkably versatile scientist, Réaumur contributed to physics (he invented a thermometer and devised a thermal scale much in use until the early twentieth century), biology (he is noted above all as the author of a twelve-volume treatise on entomology), chemistry (he investigated the composition of Chinese porcelain, demonstrated the presence of carbon in steel, and devised techniques useful in siderurgy), and other fields. To study the gastric function, he used a bird of prey as his laboratory animal. He was aware that these birds regularly eliminate in their feces the indigestible material that they take from the prey, such as bones, feathers, and so on. Taking advantage of this fact, he fed his experimental animal thin metallic tubes filled with meat. When the tubes were expelled, he could confirm that the meat was digested and had partly disappeared, while the tubes showed not the slightest mark of mechanical damage. In a further step of his research, he replaced the meat with a sponge and was thus able to collect some of the fluid secreted by the stomach, gastric juice. He demonstrated that this secretion could digest food outside the body, although quite inefficiently.

Abbot Lazzaro Spallanzani (1729–1799) took up the matter where Réaumur had left it. To demonstrate that gastric juice exerts its optimal effect when maintained at normal body temperature, he filled wooden tubes with a mixture of meat and gastric juice and affixed them under his armpits, where he carried them for several days. Then, wishing to show that the mechanism of gastric digestion is essentially the same in human beings as in animals, he swallowed small metallic tubes with many perforations, previously filled with raw meat. After failing to self-induce vomiting, he was forced to recover the tubes from his own feces. These experiments were cut short. The good abbot confessed that revulsion got the better of his determination; for once, his disgust exceeded his scientific curiosity.

In 1822, Alexis Saint-Martin, a young, strong French-Canadian trapper who was at the time in Fort Mackinac, near the Canadian border, received a musket shot in the left upper part of the abdomen. The wound was frightful: it covered "the area of a man's hand"; it affected a portion of the left lung, a fragment of which protruded through the opening; it blew out parts of two ribs; and it penetrated the stomach, whose contents poured out to the exterior. His physician saw no chance for recovery. But such were the victim's grit, endurance, and physical strength that not only did he survive but his wound healed, leaving a permanent communication (fistula) between the stomach's cavity and the outer environment.

William Beaumont (1785–1853), a military surgeon who had served in the War of 1812, took care of him. Taking advantage of the fact that the interior of the stomach was visible and easily accessible through the wound, Beaumont performed a series of experiments regarding the digestive function. He dangled a great variety of foods tied to a string into his patient's stomach, removed them at different times, observed the appearance of the mucosa, measured its temperature, collected gastric juice, and tried to correlate his observations with a number of internal and external conditions. His relationship with Alexis Saint-Martin lasted many years and was at times sour, since the patient did not take kindly to being treated as an experimental subject. In fact, on occasion he became quite angry. Beaumont's response was to observe the appearance of the gastric mucosa under the influence of emotion. His observations were consigned in a book, *Experiments and Observations on the Gastric Juice and the Physiology of Digestion* (1833).

Alexis Saint-Martin lived, always with his fistula, for another fifty-six years. He died at eighty-six years of age, on June 24, 1880, in his native Canada, having survived his doctor by twenty-seven years. His celebrity as an oddity—a living man with an open window to the stomach—he and his family found most unwelcome. It was summer when he died, and his family let his body decompose for four days before it was interred in an unmarked, concealed grave. This was done to prevent overzealous scientists from trying to secure the remains for an autopsy or some other study. Years later, a commission persuaded one of Saint-Martin's descendants to disclose the burial site, and a plaque honoring his memory was placed there. It commemorated his story, and the fact that "through his affliction he served humanity."

The study of the digestive function was further elaborated upon in the nineteenth century, when Claude Bernard established that the gastric phase does not constitute the sole aspect of digestion, which is completed in the intestine, where the action of pancreatic juice is preeminent. In this he was preceded by the work of Jean Astruc (1684–1766), who had noted the digestive action of saliva, bile, and pancreatic secretion.

As the function of the organs began to be better understood, it was possible to construct an anatomical concept of disease—that is, to attribute a disease to one organ or another. But this brought back an old epistemological problem, that of defining what is disease; for before concluding that something can be found at a certain place, it is obviously necessary to be clear about what this something is. In this regard, historians have emphasized two different conceptions of disease. One states that diseases are well-defined entities, produced by specific causes and having a distinct natural history that is proper to them—this is the "ontological" conception of disease. The second viewpoint, the "physiological" concept, posits that diseases have no reality as independent entities but are nothing other than life under quantitatively altered physiological processes. In its extreme form, the "physiological" conception of disease denied all reality to specific diseases and deemed them mere fictions constructed in order to facilitate communication and exchange of ideas, but lacking a true independent reality. Accordingly, the apparent order that resulted from classifying diseases, resting as it did on false premises, was but an "epiphenomenon."

In effect, when a man is ill, he has experiences that are uniquely his own. His suffering is of such a kind as it will never be reproduced in exactly the same form—i.e., in all its details, mental as well as physical—in any other individual. In fact, it will never be repeated in precisely the same form even in the same person at a different time. Diseases form an integral part of the life history of one individual only and are not identical to those of other individuals. This is why the "physiological" concept denied the existence of so-called disease entities. But those who observe a patient's suffering from the outside, such as his close relatives or his physician, feel compelled to try to understand what is happening to him. Therefore, they try to somehow "classify" his experience, and the first thing they do is attach a label to it, the label of "disease," which is a conceptual common denominator of many individual events. Next, the physician sees that there are certain patterns of manifestations, certain concentrations of symptoms that occur in several individuals (the European nineteenth century, with the creation of huge hospitals that permitted the examination of very large numbers of diseased persons, would make this more obvious), and the existence of common characteristics strongly suggests the idea of "disease entities."

Thomas Sydenham (1624–1689) is widely recognized as the founder of clinical medicine. Among many other contributions, he was the first to correctly distinguish between measles and scarlet fever and to identify the inheritable neurological disease known today as Sydenham's chorea. He wrote that Nature is "uniform and consistent" in the production of diseases, as a disease's symptoms are, by and large, the same in different persons. The same manifestations, as he put it, would be evinced by Socrates and by a dunce, should both happen to have the same disease. In no other way does "the universal character of a plant extend to each individual of the species." There may be hundreds, even thousands, of kinds of violets, but they are all recognizable as being violets. Statements of this kind identify Sydenham as an adherent of the "ontological" conception of disease.[10]

François Boissier de la Croix de Sauvages (1706–1767) continued Sydenham's tradition. This man, from the famed vitalist School of Montpellier (see chapter 3), took very seriously Sydenham's comparison of diseases and plants: drawing inspiration from the botanist Linnaeus (Carl von Linné, 1707–1778), he created an elaborate classification of diseases. He divided them into ten great categories in his opus *Pathologica methodica* (1759). A clear distinction between symptom and disease had not yet been established: different forms of cough (dry, accompanied by secretions, matinal, nightly, etc.) or fever (cyclical, accompanied by chills, or by sweating, etc.) were considered to be as many diseases. In consequence, his classification included the impressive number of 2,400 different disease entities.

The most illustrious member of the "ontological" school of thought was René-Théophile-Hyacinthe Laënnec (1781–1826). France was at the time a world leader in almost every field, and in medicine Laënnec, with a group of distinguished physicians, headed what became known as the School of Paris (École de Paris). Laënnec was a disciple of Jean-Nicolas Corvisart des Marets (1755–1821), Napoleon's personal physician and the first to earn the name of cardiologist, which in any case he claimed for himself. Corvisart lucidly distinguished the symptoms and signs of cardiac origin from those of pulmonary origin, a distinction theretofore clouded in confusion. Napoleon, an irredeemable skeptic, came to trust him enough to declare, "I do not believe in medicine, but I believe in Corvisart." Laënnec, like all truly brilliant students, was destined to surpass his teacher.

There were in Paris at least 20,000 hospital beds, more than in the whole of England at the time. Students from all over the world flocked to the French capital to learn from the masters of the School of Paris, whose emphasis was learning not from books but from direct observation of patients, always complemented by the autopsy findings if the patient died. Physical examination was turned into a virtuoso art. As Corvisart had done, Laënnec applied his ear directly to the patient's chest in order to discern the sounds produced by the opening and closing of the cardiac valves in health and disease, and those produced by the lung under comparable conditions. But Laënnec carried auscultation to higher perfection with his invention of the stethoscope. In one story, he was crossing the courtyard of the Louvre when he saw a boy listening with his ear to one end of a beam to hear the scratches made by a playmate at the other end of the beam. This set him thinking about the

René-Théophile-Hyacinthe Laënnec (1781–1826).

COURTESY OF THE NATIONAL LIBRARY
OF MEDICINE

possibility of applying the simple principle of sound conduction to the examination of patients. In another, more colorful version, he was to examine a particularly full-bosomed girl, when, in order not to infringe on the rules of modesty by applying his ear directly to her chest, he took a sheaf of paper, made a roll of it, placed one end on the girl's chest and his ear to the other, and was pleasantly surprised to discover that the intrathoracic sounds were transmitted to perfection.

In his *Treatise on Mediate Auscultation* (1819),[11] Laënnec described his monaural stethoscope, a wooden, cylindrical piece about nine inches long, with a detachable earpiece and chest pieces that could be screwed together. He constructed an entire system of diagnosis by auscultation, minutiously describing rales, rhonchi, whistling sounds, and the many other auscultatory findings associated with pulmonary disease. The primary focus of *Mediate Auscultation* was pulmonary tuberculosis, which was then rampant worldwide (see chapter 5) and to which Laënnec himself would eventually succumb. He noted that the tubercle (a nodular lesion characteristic of tuberculosis) was present in every organ affected with this disease. On this basis, he was able to conclude that tuberculosis was a single disease despite its bewildering range of manifestations. This was a remarkable achievement, considering that the tuberculosis bacillus had not yet been discovered and that he dispensed with the use of the microscope in his investigations. It seemed perfectly logical in his time to regard bone deformities, skin ulcers, intestinal obstruction, urinary complaints, and so on as independent, unrelated diseases. Today we know these features are a function of the various anatomical localizations of tuberculosis, but to have realized this in the premicrobial era was a stunning feat, all the more amazing when one recognizes that it was achieved by clinical observation and autopsy dissection only, unaided by modern technology.

Laënnec's genial synthesis served as inspiration to the members of the School of Paris for a series of exquisitely detailed and accurate clinical descriptions. In this way, many other diseases were identified: diphtheria and typhoid fever, by Pierre Bretonneau (1778–1862); rheumatic fever, by Jean-Baptiste Bouillaud (1796–1881); peptic ulcer, by Jean Cruveilhier (1791–1874); and others. All these physicians systematically correlated the clinical findings with the autopsy findings, thus imparting a major impulse to clinicopathological medicine. In Paris, it was easier than anywhere else to obtain clinical, hands-on experience with patients and to proceed to autopsy dissection with every death. For this reason, students flocked there from all parts of the globe.

Many foreign, Paris-trained physicians went back to their home countries, there to give unprecedented impetus to anatomical pathology. Among the most notable was the Viennese Karl von Rokitansky (1804–1878), who is said to have performed as many as 60,000 autopsy dissections in the course of his career. His work and unwavering dedication yielded many pioneering descriptions of the damage inflicted by various diseases in organs and tissues. Diseases were better understood when their characteristic ways of damaging the body were thus revealed, and so Rokitansky concluded that pathological anatomy ought to be the foundation on which the entire edifice of medicine must rest.

There was also some opposition to this manner of thinking. In France, a contrary influence was exerted by François-Joseph-Victor Broussais (1772–1838), who had been a military surgeon in Napoleon's Grande Armée. In 1808, he published his *History of Phlegmasias or Chronic Inflammations,* in which he denied the existence of any lesional specificity. To Broussais, disease entities did not exist,[12] and those who insisted on pathoanatomical correlations were perverting medicine. Disease and health were not different from each other, he argued, a stance that belonged to the "physiological" school of thought. When the body falls ill, he thought, it is because certain organs become excessively stimulated by some external irritant: cold air, foodstuffs, drugs, even moral or psychological disturbances.

Moreover, Broussais conceived the strange idea that in every case the primary focus of the irritation was the gastrointestinal system—all diseases were, at bottom, gastroenteritis, and thus, they all had to be treated the same way. Broussais, according to the practices of the time, treated them with bleeding, making him one of the most bloodthirsty physicians on record. He advocated the use of leeches, to the degree that some of his patients were covered with fifty or more leeches all over their bodies. Millions of leeches were used in France for therapeutic purposes. It was said that "Napoleon decimated France, but Broussais bled her white."

Those were tempestuous times, in which political passion was evident in every area of life. Many people were still alive who remembered the French Revolution. Revolutionary ferment, and its contrary reaction, permeated even academic life. Laënnec had received a strict religious education in a school under the aegis of the Oratorian Catholic Order. He grew up to be deeply conservative, and when he reached prominence he befriended cardinals, bishops, and former aristocrats and became the private physician of the duchess of Berry. His notes were written in Latin, even though the government had forbidden the use of that language in official teaching. In contrast, Broussais had served under Napoleon, was a thoroughgoing liberal and a materialist who extolled everything the Revolution stood for—precisely what Laënnec abhorred.

The disputes between those two formed a cause célèbre that exceeded the bounds of academic propriety and became a fight between two rancorous pamphleteers. E. H. Ackerknecht[13] wrote that not since the time of Paracelsus had anyone used such filthy, megalomaniac, fanatic, and implacable language in a medical dispute. Broussais branded his adversaries, most notably Laënnec, as "cockroaches," "poor madmen," and "men of bad faith," while their writings were "errors, lies, and vile speculations." But this fierce combatant lost the fight. In the end, his ideas became unpopular, students stopped attending his lectures, and the "physiological" concept of medicine was weakened by the emergence of a strong nosology, the classification of diseases viewed as distinct entities.

The German School served as a counterpoint to the School of Paris. Whereas the latter was centered upon hospital wards and the detailed examination of patients, the former favored scientific research, at the center of which was the laboratory. In German universities, science was all. Just as the French had enshrined the stethoscope and the minutious clinical study of sick persons, the Germans lauded the microscope and other scientific tools. This attitude produced a constellation of brilliant physician-scientists.

Justus von Liebig (1803–1873) focused on the chemistry of physiologic processes. He showed that oxidation of foodstuffs determines the body's temperature; that proteins are an integral component of living tissues, and that their degradation yields nitrogenous compounds eliminated in the urine. Even when studying infectious diseases, Liebig was more interested in the chemical changes associated with them and the mechanism whereby bacteria disrupt the chemistry of the normal organism.

Scientific research was encouraged in Germany at every level, and in every discipline brilliant men secured its conquests. Johannes Müller (1801–1858), a versatile investigator, improved the knowledge of the sensory mechanisms of vision and hearing. Carl Ludwig (1816–1895) elucidated important aspects of cardiac physiology. Hermann von Helmholtz (1821–1894) was the first to measure the speed of nervous impulses and invented the ophthalmoscope, which revolutionized ophthalmology. Jakob Henle (1809–1885) described the muscular coat of the arteries, clarified the microscopic anatomy of the eye, and discovered a part of the renal tubules that to this day bear his name, "Henle's loop." A botanist, Matthias Jakob Schleiden (1804–1881) first enunciated the cell theory, maintaining that plants were made of aggregates of living cells. Theodor Schwann (1810–1882) showed that this was also the case for animals and human beings. But it was the towering figure of Rudolf Ludwig Karl Virchow (1821–1902) that dominated medicine for most of the nineteenth century.

Virchow's life was a paragon of scholarly pursuit.[14] His productivity was not confined to medicine but was equally prolific—and no less worthy—in other fields of endeavor. At twenty-two years of age he received his doctorate from the University of Berlin. In 1848, the Prussian government assigned him to investigate a typhus epidemic in Silesia. His report went beyond medical considerations to discuss the economic, social, and hygienic factors of the outbreak; it also recommended social reforms that would prevent future epidemics. This work caused a stir among government officers on account of its unabashed democratic expressions. Virchow was attuned to the social changes and revolutionary movements that, beginning in 1848, swept across Europe. At twenty-seven he entered the political arena and was elected a member of the National Assembly by the enthusiastic populace, but he could not take his seat because he had not yet reached the parliamentary age.

He founded a medical journal in which he advocated a number of reforms in medical education and the delivery of health care. While he was still an impetuous young man, this journal became a vehicle for his political ideas, set forth with uncommon frankness. In a celebrated lampoon, he used a discussion of the heredity of disease to interject a humorous comment in which he claimed to have known of "a family, a very exalted one, in which the grandfather had softening of the brain, the son hardening of the brain, and the grandson no brains at all." Although he did not name names, everyone understood that he alluded to three kings of Prussia, Friedrich Wilhelm II, III, and IV.

The authorities were not amused by this sort of humor. In 1849, he was forced to resign his position at the Charité Hospital of Berlin. He took a job at the University of Würzburg, where he resided for seven highly productive years. Indeed, the range and volume of his writings show a titanic energy along with strong intellectual powers. He was an acknowledged leader in anthropology, ethnology, and archeological research. It was his scholarly work that guided and inspired Heinrich Schliemann (1822–1890), the excavator of Troy. Recalled to Berlin after his Würzburg years, he was elected member of the town council; during his tenure, his recommendations for public health measures transformed the city of Berlin into a model city in terms of sanitation.

By 1862, he had attained a higher position in the Prussian Chamber, but his perennial outspokenness and liberal stance set him at odds with the government. The authoritarian chancellor of Prussia, Otto von Bismarck (1815–1898), unaccustomed to seeing his views opposed, naturally became very annoyed by Virchow's stubborn contrariness. In a huff of anger, the Iron Chancellor challenged the little professor to a duel. Fortunately for medical science, this time Virchow must have reasoned that prudence was the better part of valor, and he "declined the honor." There is little question that, had the two rivals come to measure arms, the scientist would not have proven as worthy an opponent in the field as he was in the debates of the Prussian Chamber.

In the midst of all this activity, Virchow's numerous contributions to medicine were the most important. His studies of phlebitis, thrombosis, and embolism were essential to the understanding of these major medical problems. He studied infectious and parasitic diseases and wrote, among his many works, a small treatise on trichinosis. But his magnum opus was his book *Die Cellularpathologie*,[15] in which he laid out the fundamental principle that the cell is the basic unit of which all bodily organs and tissues are formed and that cells can come only from preexisting cells (hence his famous proposition, *Omnis cellula e cellula*, "Every cell [comes] from a cell"), not from a structureless matrix or "blastema," as many erroneously believed. Virchow definitively established that all diseases, "even those morbid structures which deviate most from normal structure," are to be understood in terms of cellular alterations. And whatever other agency may be involved in causing a disease (for instance, bacteria, in the case of infections), the ultimate explanation for what happens in a disease may be found in the altered vital activities of the tissue cells themselves. This fundamental principle remains the basis of scientific medicine.

As the leader of the German School, Virchow adhered to the "physiological" concept of disease. However, from our modern perspective, both the "ontological" and "physiological" concepts are now largely superseded. Classifications are acknowledged as important in medicine, yet there remains the problem of what the criterion used for classification should be. A nineteenth-century bacteriologist might have asserted that diseases should be classified according to their cause (in his field, that would mean the microorganism responsible for the disease). But even in bacteriology this would not have been an easy task. Considering the overwhelming number of bacterial families, genera, species, and subspecies, the classification would become unmanageable. There would have to be as many infectious diseases as pathogenic (and potentially pathogenic) bacterial strains. A similar objection is raised by the prestigious historian Oswei Temkin to a classification of hereditary diseases according to the gene involved.[16] And would that mean that entities such as "acute appendicitis," which every physician learns to recognize and which are immensely useful in clinical practice, would have to be discarded for the sake of a more "scientific" classification?

If the "ontological" viewpoint prevailed, it is because it is impossible to do without classifications in any field that aspires to be scientific. Of course, medical classifications, like all other concepts that denote species, are abstractions that disregard the special features of each person's experience: in classifying diseases, physicians have abstracted common qualities from the otherwise unique life experiences of sick persons. But, as a philosopher of medicine put it, "to the physician who is to live and act in the world, it is necessary to have definite categories of disease to serve as guide and tools."[17] Another reason that the classificatory approach of the "ontological" school ended up succeeding was what may be called its "therapeutic value." When we apply a name to a previously chaotic group of signs and symptoms, we make it more manageable: a certain order is created. Yet naming a threat does not abolish it. We fail to understand that attaching diagnostic labels to illnesses is a little like pronouncing voodoo exorcisms, and the apparent order artificially produced is sometimes erroneous. To which Stephen J. Kunitz, a philosopher of medicine, replies, "Whatever the case may be, affixing a name universalizes the patient's condition. It is no longer unique, and therefore no longer uniquely terrifying."[18]

Medicine today is primarily concerned with concrete problems of great scientific-technological sophistication, making the philosophical questions of former times strike us as idle or subordinate. Still, the past approaches to these questions illuminate the way in which life may be viewed at different epochs and by different societies.

In this context, a distinguished Mexican scholar, Ruy Pérez-Tamayo,[19] has expressed the opinion that the conceptions of disease in the European nineteenth century reflected the collective national personalities of the formulators. Thus, the "ontological" concept developed by the École de Paris was rational, logical, and purely theoretical. If its main premises were accepted, the conclusions imposed themselves with all the force of incontestable logic. It was very French in style: the descriptive precision of Laënnec shared the analytic spirit of Descartes and the unwavering rationalism of Voltaire. Pérez-Tamayo sees here "the spirit of immortal France, standard-bearer of the Latin genius in the Western world," and its profoundly humanistic tradition.

In contrast, the "physiological" concept of disease is above all scientific, experimental, and practical. It does not negate the usefulness of clinical pictures, but it insists that the real scientist understands that these groupings are artificial creations and not organic realities. Developed and defended by the Germanic and Anglo-Saxon peoples, the "physiological" concept is averse to adopting preconceived schemes into which each new observation must be fit. It submits to no predesigned plan but is built up gradually and tentatively from successive data provided by experience. It rejects theoretical constructs that predict and define end points: the end is known only when it is reached.

To quote Pérez-Tamayo: "The German *Volkgeist* coincides at every point with the 'physiologic' concept of disease: it is objective, realist, and definitely practical."[20]

Western medicine in the nineteenth century was being developed chiefly in France and Germany, two countries that had contrasting ways of looking at life and whose differences exploded more than once in armed confrontation. Considering that, at that time, nationalism and political ideologies permeated all human endeavors, we must conclude that medicine did not develop free of external compulsion. Science existed in a climate of national invidiousness and prejudice necessary to maintain the bellicose spirit, and medical progress more often sprang from rancorous competition between enemies than from generous collaboration between colleagues. Therefore, ascertainment of the degree to which national characteristics, real or perceived, may have influenced medical progress remains a legitimate topic for historical study.

7

THE DIAGNOSTIC PROCESS

Traditionally, the diagnostic process begins with an encounter between physician and sick person. The latter (or a third party if the patient is unable to verbalize his or her complaints, for example, young children and mentally impaired persons) describes the nature and evolution of the symptoms. This narrative constitutes the medical history as perceived by the patient. Once analyzed and interpreted by the medical practitioner, who adds the information provided by visual inspection of the patient, it becomes the "official" or "true" medical history. Up to the end of the eighteenth century, history taking and prescribing were the chief functions of the medical practitioner. Toward the middle of the nineteenth century, the physical examination started growing in importance, fostered by Laënnec's invention of the stethoscope (1816), which was adopted in English-speaking countries around 1850. Whereas before this time medical thinking had been centered on the patient's subjective impressions, now clinical diagnosis aimed at a correlation between the symptoms and signs and the morphologic findings of diseased organs revealed by autopsy studies.

In the second half of the nineteenth century, technological progress yielded tools such as X-ray machines, laryngoscopes, ophthalmoscopes, and so on, which made it possible for physicians to examine the interior of the body with unprecedented efficacy. As a result, physical examination became paramount in diagnosis and history taking of secondary importance. This trend continued until the mid–twentieth century, as further advances in biotechnology (e.g., electrocardiography, electroencephalography, and laboratory testing of patients' samples) reduced the interest in clinical history, especially among young physicians. Historians note that this attitude changed around 1940, thanks to the influence of psychoanalysis. History taking then became an interview.

The twentieth century witnessed a growing awareness of the need for international cooperation, especially in fields that, like medicine, seek to relieve suffering without regard to age, sex, social condition, or ethnic or national origin. Physicians practicing the "art of diagnosis" drew from all useful venues, and the process became the same, in its general outlines, as it is today. History taking is presently regarded as a very important first stage in making a diagnosis. This is a task that requires skill, tact, discriminatory sense, and receptive intelligence on the part of the physician, who must be alert to nonverbal clues, which sometimes convey more information than the patient's actual complaints. Ideally, the patient's history includes the family history, the individual's past history, and the psychosocial and emotional factors that often color the medical condition. Needless to say, a skilled interviewer must establish a true rapport in order to be successful. The modern physician has also inherited a method of questioning patients. A review of major bodily systems (usually starting at the head and proceeding downward) is aimed at avoiding overlooking any past or present problems that may be influencing the patient's health.

Next follows the physical examination, during which the physician relies on his or her senses. Visual inspection with the unaided eye reveals the patient's general appearance (i.e., bodily symmetry, nutritional status, posture). It is easy to get lost in individual details and to fail to see informative aspects of the overall appearance (personal habits reflected in grooming, clothing, etc.). Hippocrates attributed great importance to the appearance of the face and the eyes, which indicated the gravity of the disease (Hippocratic facies), and to the position of the patient in bed, which reflected the intensity of the suffering. The physician also looks for specific features of diagnostic value. Thus, "clubbing" of the nails (broadened, concave, shiny nails, already mentioned and correctly interpreted in Hippocratic texts) bespeak prolonged defective tissue oxygenation, as with congenital heart disease or obstructive pulmonary emphysema; pale mucosae and conjunctivae indicate anemia; bronze coloration may reflect abnormal iron storage; some bodily anomalies indicate heritable syndromes (for instance, a skin fold on the inner angle of the eye and a transverse crease on the palm of the hand are signs of Down syndrome); and so on.

The greatest revolution in visual diagnosis started in 1895, when a modest German physicist, Wilhelm Conrad Röntgen (1845–1923), accidentally discovered that a species of rays theretofore unknown (hence named "X-rays"), emerging from a cathode ray tube, could pass through many objects, including the structures of the human body, and leave an image on a photographic plate or a fluorescent screen. The first radiograph taken, that of Röntgen's wife's hand, promptly went around the world. This extraordinary discovery, truly one of the greatest benefits to mankind, did not alter Röntgen's modest, quiet disposition. He resisted the many exhortations to patent his invention, which would have made him immensely rich, on the ground that discoveries that benefit the entire human race must be available to all.

Röntgen's invention was soon replicated in many parts of the world (the initial apparatus, a pear-shaped cathode ray tube made of glass, was common in many physics laboratories, inexpensive, and of simple design). Anyone who wished to copy the invention and could afford it actually did so. In the United States, a veritable "roentgen craze" seized the country: there were X-ray Boys Clubs, and coin-operated machines, where everyone could have his own roentgenogram taken for a low price, were installed in several large cities.[1] This bounteous enthusiasm would cool when it became apparent that injudiciously applied X-rays were the cause of severe burns and terrible complications.

At the end of the nineteenth century and the beginning of the twentieth, there was great interest in spiritualism, animal magnetism, extrasensory perception, otherworldly communication, and the like. Photographic plates were used in naive or deceitful attempts to catch the image of ghosts and spectral presences. It was only natural that roentgen rays should enter into this preoccupation. Religious sentiment had weakened, and the spiritual vacuum it left induced many people to hope that some form of unperceived reality, a transcending

Caricature published in 1897, shortly after the discovery of X-rays, when the innovation had greatly excited the popular imagination. The caption reads: "Whether stout or thin, X-rays make the whole world kin."

universe, would be revealed by science. Many persons, including reputable scientists, set out to explore the metaphysical aspects of these new rays. This attitude was further strengthened when Marie Skłodowska Curie (1867–1934) and her husband, Pierre Curie (1859–1906), discovered yet another new form of energy, radioactivity.

But there were other consequences of the discovery of X-rays that went beyond the confines of medicine. The lives of women in society were strongly impacted. Until then, women's bodies had been zealously guarded from strangers' glances. A possessive patriarchy regarded the woman's body as part of its property rights. Although a veil hiding the face was never in fashion, puritanical morality effectively veiled most of the female anatomy. Physicians in the nineteenth century dared not look at their female patients, who were sometimes placed behind thick curtains, through which they extended a limited part of their body to the examiner's inspection. The vaginal speculum, although freely used on the European continent, was adamantly rejected in North America as outrageously immoral.[2] And even during parturition, there is evidence that the recommended position for childbirth obeyed more the dictates of modesty and conventional morality than any sound medical reason.

In this repressive social climate, the fact that a viewer could look past a woman's clothing stirred the deepest layers of prejudice. There were excesses of ludicrous nonsense: in one alleged episode, a New Jersey congressman tried to introduce a law barring the use of X-ray binoculars, and a London clothing store marketed X-ray–proof, presumably lead-lined, underwear.[3] It was primarily the skeleton that roentgenology disclosed. But the new visibility of the woman's body also forced a reconsideration of repressive attitudes. It led women to question many of the practices that had kept them secluded. It exposed the skeletal deformities inflicted by the irrational custom of wearing overly tight corsets. Women acquired a new sense of power, enhanced knowledge of healthy practices, and greater consciousness of their own bodies.

Once radiography came into widespread use, the American physiologist Walter Bradford Cannon (1871–1945) made it possible to visualize viscera that normally could not be seen by X-rays using bismuth salts or barium sulfate as contrast media. Walter B. Cannon was one of the most distinguished physiologists of the twentieth century. His book *Autonomic Effector Systems,* written in collaboration with the Mexican scientist Arturo Rosenblueth (1900–1970), was a major landmark in the study of the autonomic nervous system. Cannon's mentor at Harvard, Henry Pickering Bowditch (1840–1911), had in turn been a student of Carl Ludwig, in Leipzig, symbolizing the fact that the leadership of biomedical science had now passed to the United States of America. In the twentieth century, this country acquired the undisputed supremacy in medicine and the biological sciences.

In our day, highly developed technologies permit us to visualize the internal structures of the body, each one yielding its particular kind of diagnostic information. Computerized tomography, commonly known as CT (also CAT, the "A" standing for "axial"), was invented in the 1970s. A British engineer, Godfrey Hounsfield, and an American physicist, Allan Cormack, are commonly acknowledged as the inventors. This method would not have been possible without the major advances that had taken place in the field of computers. CT uses parallel X-rays that go through the body, but, unlike conventional radiography, do not directly make an impression on a photographic plate; instead, the rays excite many sensors that send signals to a computer, where they are processed and turned into pixels on a video monitor. CT produces pictures of cross sections or "slices" of the interior of the body of extraordinary resolution: blood vessels, bones, and soft tissues of various densities can be clearly distinguished; it is one of the best tools for the study of the chest and abdomen. The high resolution of the images obtained facilitates the study of delicate and complex structures such as the eye and the inner ear. Refinements such as the use of contrast media or the colorization of the digital images obtained increase the diagnostic usefulness of this technique.

Nuclear magnetic resonance imaging, usually abbreviated MRI (the word "nuclear" was dropped from common usage because of the negative connotations it carries among the public), came after CT; its use was first authorized in the United States in 1984. MRI benefited greatly from the remarkable progress in computer technology that CT had instigated. The 2003 Nobel Prize in Physiology or Medicine was awarded to Paul Lauterbur (b. 1929), from the United States, and Peter Mansfield (b. 1933), from England, for their work on MRI. This method is based on the physics phenomenon that when certain natural elements are subjected to a strong magnetic field under certain specified conditions, their atomic nuclei give off energy signals that carry an encoded message of the physical and chemical characteristics of the environment. This phenomenon applies as well to the human body, in which hydrogen, as a component of water, is the most abundant element responsible for the signal. Advanced computer technology decodes the message and transforms it into images of exquisite resolution. Therefore, during MRI, a patient is not exposed to potentially harmful X-rays, only to a powerful magnetic field, and no harmful effects have been reported as a result of this exposure.

Anatomical structures composed of cartilage, bone, and soft tissue, such as the joints, are seen much better using MRI than CT, and the central nervous system, encased within the skeleton, is revealed by MRI with astounding clarity and detail. Like other imaging techniques, MRI is constantly being perfected. Methods now exist of obtaining three-dimensional images that are invaluable in localizing lesions. Another improvement, "fast" MRI (fMRI), captures rapid-flowing images, enabling us to see living organs during their biological or metabolic activity. Brain function, for instance, is known to be associated with increased blood flow. Since the magnetic signal of oxygenated arterial blood is different from that of deoxygenated venous blood, MRI allows the detection of cerebral activity.

In this connection, astonishing breakthroughs have come from the technique known as positron emission tomography (PET). Positrons are positively charged subatomic particles emitted by isotopes. As these particles travel out from the nucleus, they collide with electrons, producing gamma rays, which, detected by the sensors of a PET machine, are transmitted to computers able to reconstruct a picture of the area of the body that emitted the radiation. Although the image thus produced is a "tomogram," i.e., a "slice" or plane of the body, it is not meant, like CT or MRI, to show the fine details of the anatomical structures represented. Instead, PET aims to reveal certain functional aspects of the body and their alterations in disease. A radioisotope is administered (ingested, injected, or inhaled) to the patient. Once absorbed, it concentrates in certain organs of the body (for instance, radioiodine is taken up preferentially by the thyroid gland). In this manner, the images obtained by PET scanning allow the measurement of the local concentration of the substance (reflected by the intensity of the signal) and its distribution in the body. When the radioactive compound is attached to glucose, it can be used as a tracer to study cerebral blood flow, because it goes to the areas that most need the energy furnished by glucose. In this manner, it is possible to study the way in which the brain utilizes energy while doing mathematical calculations, solving puzzles, watching visual stimuli, or engaging in other forms of mental activity. Thus, imaging techniques such as PET and MRI have made it possible for the first time to "see the brain thinking," that is, to scientifically analyze objective changes linked to mental activity.

Since malignant tumors concentrate certain isotopes, tracking the development of metastases from malignant tumors is greatly facilitated, sometimes even before they become visible by any other technique. The radioisotopes used decay very fast, and the low doses used pose no danger to the patients. These techniques are only part of the now-vast field of nuclear medicine.

CT, MRI, and PET require elaborate, complex, and costly technology. They are available only at medical centers that have the appropriate construction and expert staff. In comparison, ultrasound imaging is relatively simple. A transducer emits sound waves that pass through the body and encounter structures of varying sound conductivity. The sound waves bounce back to the transducer, and a computer interprets the echoes in terms of the size, shape, and distance of the objects encountered, thus producing an image. The use of sound waves to detect the shape of objects goes back to the nineteenth century and received great impetus during the two world wars, when the concern was the detection of enemy submarines. Attempts to apply this technique to the human body began in Austria during the 1930s and were followed after World War II by the efforts of numerous researchers, who solved diverse technical problems and made its medical application possible.

Ultrasound machines for clinical use became widely available after 1975, when it was possible to obtain "real-time" images, particularly clear, undistorted pictures of moving structures. Nephrologists and urologists found this technique particularly helpful in diagnosis, but it was obstetricians who became the major users, largely because social changes occurring since the late 1960s tended to empower women, who then desired to "see for themselves" the prenatal development of their children, instead of relying on the word of medical experts, most of them men. Since then, ultrasound diagnosis has become a standard method in obstetrics.

To the already described methods that enable physicians to look into the interior of the body must be added instrumental screening by means of flexible fiber-optic instruments to directly visualize and photograph the inside of hollow viscera and bodily cavities (sigmoidoscopy, nasopharyngolaryngoscopy, gastroesophagoscopy, laparoscopy, etc.).

———

Next in the physical examination comes the collection of auditory information. This is done by direct auscultation with a stethoscope of the sounds generated in the chest and abdomen. Much literature on the subject has accumulated since the times of Laënnec, and the modern student of medicine depends on sound recordings that, better than any verbal description, demonstrate what is meant by rales, crackles, wheezes, murmurs, and sundry auditory phenomena helpful in diagnosis. The ancient Greeks practiced auscultation of the chest, which apparently was later forgotten. Although Laënnec, who was quite familiar with the Hippocratic texts, must have known this, he scarcely acknowledged it.

Auditory data are also gathered by percussing the chest and abdomen. There is evidence, contained in the Ebers Papyrus (numbers 189 and 864),[4] that percussion as a diagnostic means was used by ancient Egyptian physicians. In modern times, the method of chest percussion was first described by Leopold Auenbrugger von Auenbrugg (1722–1809), an Austrian physician born in Graz, in an inn where his father used to strike barrels in order to determine the level of liquid inside. Presumably, this was the source of his idea to percuss the chests of patients when he became a physician working in the Spanish military hospital of Vienna. There, he made the important observation that when the thorax of a healthy individual, not obese, is percussed, a resonant sound results. The quality of the sound is altered when there is underlying pathology. At first, he used to strike the clothed chest with the tips of all fingers held together or using a leather glove when striking the bare skin. This was later replaced by mediate percussion, in which the fingers (or a small hammer) struck a resonant solid body, called a "plessimeter," held against the surface of the chest. Later still, it became more practical to apply the hand (usually the left) against the region whose sonority is being investigated, and tap with short, sharp blows one to three fingers with the fingers of the other hand (digitodigital percussion). Auenbrugger was very fond of music and well versed in this art, and this must have predisposed him to develop an audition-dependent method. He even wrote the libretto for the comic opera *Der Rauchfangkehrer* ("The Chimney Sweep"), composed by Antonio Salieri (1750–1825).

Auenbrugger tried his method on numerous living patients and on cadavers. The latter allowed him to confirm the validity of his findings: a normal lung is resonant to percussion; an emphysematous lung, containing more air, is hyperresonant; solidification of lung tissue produces a dull sound, which may be present over a localized area, as in the case of a tumor or some discrete mass lesion, or generalized, as in pneumonia or pleural effusion. Although Auenbrugger published his findings in a book in 1761, percussion met with skepticism and did not become widely adopted until it was endorsed by the famous Corvisart nearly forty years later.

The physician also uses touch, i.e., palpation, as a diagnostic aid. In the Chinese medical tradition, touching the pulse had a cardinal significance. It may be said that it acquired the status of a science, "sphygmology." In the oldest known Chinese medical text, the *Huang di nei jing* (The Yellow Emperor's Classic of Internal Medicine), pulse taking is the chief means of diagnosing diseases; all others are subsidiary.[5] Taking the pulse was a very complicated process: there were rules governing which season of the year, and which part of the day, were optimal for assessing it. Even the sex of the patient influenced the procedure: if it was a woman, the left pulses were taken first; if a man, the right pulses. "Pulses," in the plural, is a good distinction, because the physician applied three fingers and tried to distinguish the characteristic sensation in each of the palpating fingers. At least twenty-eight different kinds of pulse, with many variants, could be discerned by palpation, and these were informative of which acupuncture points might be effectively stimulated. Acupuncture tracts and sphygmology were inextricably joined to each other.

A seventeenth-century English scholar wrote:

> The Circulation of the Blood, which with us is a modern Discovery, has been known there, according to Vossius, four thousand years; they have such a skill in Pulses as is not to be imagin'd, but by those who are acquainted with them.... Even the Missionaries, who have reason to know them best, grant, that there is somewhat surprising in their Skill of Pulses [and] tell us that they have made Observations in Medicine four thousand years.[6]

It seems that the Chinese were more concerned with the pulse's *qualities* than with its frequency or rhythm. But this does not mean that the latter features were neglected: the Chinese physician learned to take his own pulse in order to compare it with the patient's. In the West, physicians thought that different temperaments were revealed by the pulse, and in the seventeenth century they devised baroque classifications of the different kinds of pulse that presumably they were able to detect, but their pulse taking never attained the degree of sophistication to which it was raised by their Chinese counterparts. Instead, in the West the notion of quantitation seems to have predominated from very early on. Erasistratus of Ceos (see chapter 1) constructed a special water clock (clepsydra), specifically devised to measure the pulse.

The modern physician inherited the knowledge that by touching the surface of the body lightly or deeply, information may be gained concerning the state of the organs beneath the surface. Lesions may be detected by palpation in breasts, prostate, testicles, or lymph nodes. Deep palpation of the abdomen detects abnormalities in liver, kidneys, or spleen. Light palpation elicits abdominal muscular spasm in cases of peritonitis. Palpation may also detect a vibratory sensation in the chest ("thrill"), produced by an area of blood turbulence when there is an abnormality of the cardiac valves. Historically, however, diagnostic touching was looked at askance in the past, especially when it involved contact with a person of the opposite sex. This may have been one reason why physicians turned to an examination of the patient's secretions and excretions as important adjuncts in diagnosis.

For the examination of bodily fluids, the modern physician relies on the analytical laboratory. Here the most frequently performed analysis is that of urine, the chemical composition of which yields crucial information for diagnosis and prognosis. But before laboratory medicine existed, physicians practiced the examination of urine by the senses. This was "uroscopy," a medical practice that in the past was tainted by quackery, as described in a number of publications.[7] Since at least the twelfth century, physicians have routinely examined the urine when consulting on a patient, but an Ostrogothic document from the sixth century already says that the urine "tells a skilled physician the whole history of his patient's disease."[8] Illustrated medieval manuscripts often show the physician looking at a urine-containing flask made of clear glass of rounded bottom, called a *matula*, which he holds up against the light. As late as the seventeenth century, more than one unscrupulous physician pretended to diagnose and prognosticate without looking at the patient, simply by inspecting the urine.

What did they see? A number of important characteristics depending on the color, odor, fluidity, limpidity, nature, and "hypostasis" (meaning sediments) of the urine. For instance, orange-colored urine indicated good health; saffron-colored, the presence of bile; red, blood and therefore a bad prognosis. But a great variety of hues came to be recognized: green, brown, violaceous, livid, and so on, whose mere enumeration would be tedious here. The urine's foam had to be carefully noted. If abundant, the patient was prone to suffering from colic, due to excessive air in the viscera. If the bubbles were small, this indicated a proclivity to migraine. Yet not all "uroscopy" was pure nonsense. Hippocrates, in the *Aphorisms,* had noted that foamy urine is associated with chronic diseases. Modern commentators remark that bubbles on the surface of the urine may indicate an abnormal concentration of protein, a mark of abnormal renal function. By the same token, a saffron color betrays the presence of the bile pigments bilirubin and urobilinogen. In the Middle Ages, Avicenna noted the offensive odor of the urine in some patients with chronic diseases and its ability to stain linen. The offensive odor is now known to be due to certain chemicals (mercaptans) present in cirrhosis of the liver,[9] just as the linen-staining property relates to the presence of bile pigments.

In sum, before physicians could rely on the laboratory, they did what they could in the area of diagnosis. Considering the rudimentary state of biomedical science and the highly subjective nature of sense perception, this was not much. Moreover, subjective methods lend themselves to quackery and imposture, which, sadly, were prevalent in prescientific days. But a few brilliant observations were made, whose enduring value attests to the sagacity of the clinicians of yore. "One is stunned by the clinical sense of our precursors," exclaimed one prestigious twentieth-century physician.

A number of artistic renderings depict the physician in the act of examining the urine. The sardonic genius of William Hogarth (1697–1764) came up with a specially caustic representation. His engraving entitled "The Company of Undertakers" was designed as a coat of arms for the medical profession. In the upper part he represented three personages who presumably were caricatures of renowned contemporary doctors. The central personage, an ophthalmologist, is cross-eyed. In the lower part of the engraving a dozen doctors coiffed with monumental wigs huddle together. They all carry canes, whose heads they raise toward their noses (in the eighteenth century the head of a doctor's cane contained aromatic substances, supposedly to preserve him from contagion). Two medicos are looking at a urinal, to evaluate the characteristics of its contents. A third one is dipping his fingers in it, in order to taste the urine. This, indeed, was done by conscientious medical men.

The modern physician has plenty of reasons to be thankful that he, or she, does not have to rely on his senses as much as his forerunners did. Occasionally articles appear in the medical literature lamenting that medical students today do not know the valuable diagnostic information that can be conveyed by the sense of smell, since physicians of the past knew that smell was clinically useful (for instance, the urine of

William Hogarth's "The Company of Undertakers."

patients with phenylketonuria has a musty, barny scent; that of those with scrofula has the odor of stale beer; patients in diabetic coma emit an acid-fruity aroma; and so on) and saved time and money by pointing immediately to the right diagnosis. But the loss of this skill seems less regrettable to those who consider the common practices antedating the laboratory age. Then the physician did not simply ask the patient how his stools were but actually looked attentively into the chamber pot and closely approached it to his nose, the better to catch the emanations: there was much to learn from a patient's dejecta.

Documents show that in the seventeenth century the characteristics of the stools of the great personages of history were duly recorded by their physicians. Pierre Bourdelot, private physician to the duke of Bourbon (1692–1740), grandson of the Great Condé, kept a careful record of his exalted patient's bowel movements. A note reads, "He went only three times to the commode, but all the [fecal] matters, accumulated for a long time at the bottom of the mesentery, and burnt by their sojourn, were of such color as if he were twenty years old." And Dr. Guy-Crescent Fagot (1638–1718), physician in chief at the court of Louis XIV, kept a daily journal of the health status of his royal patient, the "Journal de la santé du roi," thanks to which we know more than anyone could possibly care to know about the king's excreta. According to it, the Sun King spent a great deal of time sitting on the *chaise-percée,* the night commode. The journal records, day by day, the most minor details referable to the monarch's health. An example:

> On Sunday, the ninth of this month, the king went ten times to the commode, from his rising from bed to four o'clock in the afternoon, when feeling tired he chose to rest in bed, and fell asleep until nine o'clock. Upon waking up he went again to the commode, releasing raw and undigested matters, and, having wished to take only a tincture of sage and veronica instead of foods, he went back to sleep.... But thereafter his loose stools mixed with undigested matter have continued to run, and the king was obligated to repose in bed, and there to hear Mass...sticking to the regime that I was honored to propose to him.[10]

As science and technology progressed, physicians were freed from the need to expose themselves to disagreeable sensations. Laboratory instruments could now gather objective information that was more consistent and accurate than any obtained through the subjective sense impressions of a physician. The greatest impetus to clinical diagnosis by laboratory means came when microscopy reinforced chemical analysis. "Uroscopy" became "urinalysis." No longer was it necessary, in order to diagnose a urinary tract infection, to look at the urine's turbidity, to smell its repellent odor, or, as some had done, to taste it! It sufficed now to detect chemically the presence of nitrites and to observe the leukocytes in the sediment (normally, up to five white blood cells are present per high-power field; more than ten indicate infection). By the same token, a number of structures are revealed by microscopy in the urinary sediment that are helpful to diagnosis, such as casts made of cells and protein material, red blood cells, crystals, and so on.

Microscopy complements and expands the techniques of bacteriology and parasitology. It affords the unique intellectual satisfaction that comes with directly visualizing the cause of a disease or the structural abnormality responsible for the symptoms. An example of the former is the detection of eggs of parasites in the stools or of infection-causing fungi; as to structural abnormality, no better example may be quoted than the ability of microscopy to detect the cellular changes of cancer or its precursor lesions.

As important as microscopy is to diagnosis, historians wonder why it was not systematically applied to medical problems much earlier than it was, in the late nineteenth century. In effect, the microscope and its wondrous ability to reveal an unsuspected "universe of marvels" were widely known from the 1600s. But it took nearly 250 years before it became the indispensable tool of a medical diagnostic workup.

One reasonable explanation for this lag was provided by Drs. Guido Majno and Isabelle Joris.[11] Dr. Majno practices what he calls "experimental history"—he re-creates the exact conditions of the past and tests hoary claims allegedly based on those conditions with up-to-date technology. For instance, if the Hippocratic texts recommend using the juice of the fig tree to stanch bleeding, Dr. Majno will apply known amounts of fig tree latex to blood samples and other coagulable substances and then try to determine if blood coagulation is promoted, as measured by today's coagulation tests.[12] (It is not, but fig tree juice curdles milk, and thus the Greeks reasoned by analogy without the benefit of experimentation.) To address the late advent of medical microscopy, Majno and Joris used a Culpeper-type microscope constructed before 1799 and photographed some of the blurry images obtained with this instrument. They concluded that the optics of early microscopes were badly deficient and the distorted, unreliable images had been more a hindrance than a help to researchers of that era. The instruments were also very expensive: only wealthy aristocrats could afford them, and they tended to think of the microscope as more of a toy for their personal amusement than a working tool for serious-minded medical scientists.

Another reason the microscope took so long to become popular was that no one knew how to make use of it. Anatomy had been thoroughly explored by the long and laborious efforts of dissectionists; physiology had begun to make significant inroads. But there was still no clear idea about what happened to the organs during disease. Moreover, techniques permitting the ascertainment of the microscopic constitution of bodily structures had yet to be developed. Microscopy was only a technique, a means of elaborating, refining, and perfecting the available *anatomical* knowledge, which would thus be carried to levels of detail previously undreamed of. But before the *medical* knowledge could be similarly improved, there had to be a clear concept of precisely what was to be refined and perfected. In other words, there had to be a conceptual framework of *pathological* anatomy. Here, ideas were still blurred and inchoate, although there had been no lack of diligent and brilliant precursors.

In the Middle Ages, cadavers were occasionally opened to explain some aspect of a person's life that was thought especially puzzling. This could be a sudden and unexplained death or suspicion of poisoning or other foul play. Sometimes the reason for the autopsy was neither medical nor forensic. For instance, the corpse of a Church member noted for piety, faith, and self-abnegation might be "opened" in order to find evidence of sainthood or to secure relics. In those cases, it could scarcely be said that the autopsy yielded any medically useful information. Performed by ardent coreligionists, any finding was likely to be interpreted with a strong theological bias. For instance, a gallbladder containing three large calculi found in the body of a revered clergyman was deemed a mark of the Holy Trinity. Gross pathology is a visual, and therefore subjective, activity. As we have discussed, observers tend to see only what they wish to see or what they have been conditioned to see.[13]

The anatomists who performed dissections in the Renaissance must sometimes have seen diseased organs. Indeed, we have seen that Matteo Realdo Colombo (1516?–1559), a successor of Vesalius, actually reserved Book XV of his main opus, *De re anatomica,* for a description of the anomalies and curiosities that he had come across in the course of his career. But an attempt to understand the lesions encountered, and to systematize the observations, was made by only a few brilliant minds. Especially noteworthy is Antonio Benivieni (1443–1502), a cultured Florentine physician and friend of the most distinguished artists and intellectuals of his age, including the philosopher Marsilio Ficino (1433–1499), the poet Angelo Poliziano (1454–1494), and the fiery priest and political reformer Girolamo Savonarola (1452–1498). The last was among his patients, together with members of the prominent Medici, Benci, and Guicciardini families. Benivieni wrote a brief book entitled *On Some Occult and Marvelous Causes of Diseases and Their Cures* (*De abditis nonnullis ac mirandis morborum et sanationum causis*),[14] published five years after his death but, remarkably, seven years before the birth of Vesalius. It is a fascinating book that offers an insight into the lives of people in Renaissance Florence, because their social and clinical histories are recorded together with the necropsy protocols.

De abditis is widely regarded as the first book of anatomical pathology. From a few of its case descriptions, the modern pathologist can make a retrospective diagnosis. But most clinical histories are skimpy, because the book's goal was eminently practical; and the recorded data are so unlike today's as to make diagnosis nearly impossible. Laudable as this effort was—one of the first to try to correlate clinical observations with the postmortem findings—it did not revolutionize the existing concepts of disease. Benivieni was very much a man of his time, and he subordinated all his observations and interpretations to the Galenic theory of the four humors and other prevailing, erroneous notions.

Théophile Bonet (1620–1689) was another important student of pathological anatomy. This Swiss physician was an indefatigable compiler. In 1679 he published his most important work, a ponderous treatise in three volumes and 1,706 pages, within which is consigned all the experience that anatomical pathologists had been able to accumulate thereto. The descriptions include 2,934 clinical cases and their respective autopsies, as well as quotations from the experience of 407 authors. He chose to name his work *Sepulchretum,* meaning "cemetery," since it was, in a sense, a repository of cadavers. Its very long and bombastic complete title was worthy of the monumental compilation that it heads. Approximately, it was "A cemetery or anatomy of cadavers dead of disease, wherein are communicated the histories and observations on all the alterations of the human body, and their occult causes are revealed. In fact, it deserves to be called the foundation of the true pathology, and of the appropriate treatment of disease, and even the inspiration of medicine, ancient and recent" (*Sepulchretum, sive, anatomia practica ex cadaveribus morbo denatis: proponent historias et observationes omnium pene humani corporis affectuum, ipsorumq[ue]; causas reconditas revelans. Quo nomine, tam pathologiae genuinae, quam nosocomiae orthodoxae fundatrix, imo medicinae veteris ac novae promptuarium dici meretur*).

Some say that Bonet put together his enormous compilation with little discrimination, mixing the important with the trivial and displaying a morbid proclivity to dwell on human malformations or monstrosities. Such an assessment is unfair. In the first edition of *Sepulchretum*, published in 1679, there was a painstaking effort to arrange the protocols in systematic fashion and to facilitate consultation with numerous cross-indexes. These disappeared from the second edition (1700), when the editor adjudicated them unnecessary. Recent scholarship considers Bonet one of the founders of anatomical pathology,[15] whose laudable patience and impressive erudition focused the attention of the medical community on the importance of anatomoclinical correlations.

Bonet's work fostered the advent of one of the towering figures of pathology, Giovanni Battista Morgagni (1682–1771). Born in Forlì, near Bologna, where he studied medicine, Morgagni was the first modern pathologist; with him this discipline became a true science. Contrary to his predecessors, or even to his most renowned contemporaries, such as the genial Hermann Boerhaave (1668–1734), the modern founder of bedside clinical teaching, Morgagni did not perform his postmortem examinations solely to find the cause of death. He was patiently constructing a whole theoretical edifice, an elaborate conceptual framework, which was central to pathology becoming a true scientific discipline. His main opus was entitled *On the Seats and Causes of Disease, Investigated by Anatomy* (*De sedibus et causis morborum per anatomen indagatis*). Written in the form of seventy letters addressed to his friends, the book describes more than seven hundred clinical cases and their respective autopsy reports with exemplary minuteness. And in every case the comments of Morgagni aim to correlate the clinical manifestations with the pathological changes uncovered by postmortem study. From then on, the patient's symptoms were inextricably linked, in the physi-

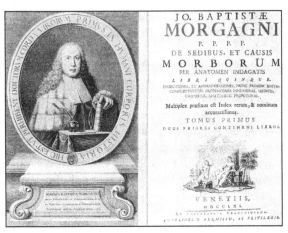

Giovanni Battista Morgagni (1682–1771) and the title page of his main work, On the Seats and Causes of Disease, Investigated by Anatomy.

cian's mind, to structural abnormality. The "anatomoclinical method" had been born, and it would prove immensely fruitful in medicine.

The first phase of modernity in medicine, which Morgagni instituted, continued through the nineteenth century. Careful observations led him to identify a number of specific pathological conditions: syphilitic aortic aneurysms, endocarditis, mitral stenosis, regional ileitis, ovarian and paraovarian cysts (the latter known to this day as "hydatids of Morgagni"), cirrhosis of the liver, cancer of the stomach, gastric ulcer, cerebrovascular accidents, and many others. The title of his book referred to "the seats of disease," and he proved irrefutably that diseases had a primary localization. If any vestiges still remained of the old Galenic concept that viewed disease as a generalized disturbance in the equilibrium of the "humors," this certainly was demolished once and for all by the work of "His Anatomical Majesty," as Morgagni's students called him.

But with his book, Morgagni claimed to demonstrate not only "the seat" (*sedibus*), but also the causes of diseases (*causis morborum*). This was not the case. One does not find the cause of a destructive phenomenon merely by looking at the wreck it leaves behind, however prolix and assiduous an observer one may be. The investigation of causes necessitates a dynamic approach, which only well-designed experiments can provide. Still, the suggestive power of morphology joined to clinical observation is so strong that it has often determined the direction of experimentation. Thus, if a history of alcoholism is repeatedly associated with a cirrhotic liver, the mere association, no matter how reiterative, does not constitute proof that the two are related as cause and effect; but it certainly suggests that experiments designed to explore the hepatic toxicity of alcohol promise to be fruitful.

The anatomoclinical method inaugurated by Morgagni received a marked impulse in the nineteenth century by Xavier Bichat (1771–1802), whom we have previously encountered as a leading "vitalist" (see chapter 3). Bichat believed that "to perform dissections in anatomy and experiments in physiology, to follow the patients and to open cadavers in medicine: such is the triple path out of which there can be neither anatomist, nor physiologist, nor physician." With Bichat, medicine began a process of evolution that led to our contemporary health science. He shifted pathology from organs to tissues, an amazing accomplishment when one considers that it was done without using the microscope. His unexcelled observational powers, coupled with his experience in dissection and a fertile mind, converged to give him the idea that organs were formed of "membranes," as he called them. These anatomical elements could be of the same kind in different organs, reacted in the same way, and, he proposed, were in some cases separable from the organ to which they belonged (he isolated them by teasing, applying heat or acids, and employing other procedures). These "membranes" are now called "tissues." Bichat described twenty-one different kinds, some of which are conceptually as useful today as when he first identified them (serous, synovial, cartilaginous, medullary, fibrous, etc.).

Although Morgagni and Bichat had not used the microscope, this instrument was masterfully exploited by German investigators. Thanks to their efforts a cellular theory was formulated, primarily by Theodor Schwann (1810–1882) and Matthias Jakob Schleiden (1804–1881). The chief postulate is that "all living organisms are constructed according to the same principle, and this principle is the formation of cells."[16] This is a law of universal application in biology that is comparable, in its all-encompassing breadth and unerring validity, to the law of universal gravitation in physics or to the second law of thermodynamics. And it is fitting that its formulators should have been a zoologist (Schwann) and a botanist (Schleiden). But its most ardent defender was a physician and scholar, the incomparable Rudolf Virchow, to whom we have referred in some detail before (chapter 6).

For some time during the nineteenth and twentieth centuries, the ultimate basis of diagnosis was morphology; in other words, pathologic anatomy. Diseases were conceptualized as departures from normal structure. The pathologist founded his diagnosis on a principle of analogy: normal and abnormal tissues were compared, and the morphological variation between one and the other was the defining criterion of a lesion. A lesion was a morphological alteration detectable by any valid observational technique (usually microscopy). According to the cellular theory, elementary lesions occur at the cellular level; and these elementary lesions aggregate in distinctive patterns that allow for a diagnosis. When these changes are confronted with the clinical manifestations (the anatomoclinical method propounded since Morgagni), the idea emerges of a "pathologic entity." This is how nosology, the classification of diseases, came into its own in the modern era; most of the diseases that we know today were discovered in this fashion. And relying on observations of the progress of other similar cases, one could begin to formulate a prognosis.

Since the 1950s, pure morphology ceased being the sole criterion for diagnosis. A number of new technologies were added to the traditional methods, such as electron microscopy, histochemistry, immunohistochemistry, flow cytometry, cytogenetics, and others. The discovery of DNA opened new vistas. Its magnificent saga—which started with the isolation of "nuclein" by Johann Friedrich Miescher (1844–1895) and culminated with the much-chronicled ascertainment of its molecular structure by the combined efforts of many investigators, notably Maurice H. F. Wilkins (1916–2004), Rosalind Franklin (1920–1958), James Watson (b. 1928), and Francis Crick (1916–2004)—cannot be adequately covered here. This has been called the most important discovery in biology in the last hundred years and has been abundantly documented. DNA is revolutionizing diagnosis, where considerations now shift from the cellular to the molecular level. "All biologic phenomena are engineered at the molecular level," declared one of the field's pioneers, "and we will not understand life unless we understand the interactions of molecules."

Although the contribution of morphology to diagnosis is still crucial, it is no longer its sole organizing principle. For instance, the morphological diagnosis of some malignant tumors is no longer considered valid unless it takes into consideration the findings of flow cytometry, immunohistochemistry, and molecular biology. The last discipline has evolved to a remarkable degree of sensitivity. DNA probes now exist that can scan thousands of genes and pick up minute genetic abnormalities, down to those affecting a single base-pair change, among the three billion base pairs that make up the entire human genome.

But for all the mind-boggling sophistication of modern diagnostic technologies, the diagnostic process still demands the physician's prudence and good judgment. Tests may be extremely sensitive but are not necessarily specific to the diagnosis of a given disease; very rarely has a diagnosis been based only on test results. Ordering too many tests may be confusing—not to say costly—since it often imposes the need for further confirmatory testing. On occasion, spurious abnormal results demand more testing to confirm or discard them, with the discomfort and greater financial burden this means to the patient—a consequence all the more lamentable if the tests were not indicated in the first place. Thus, the physician must know what tests to order and when to request them.

As in the days of Hippocrates, the diagnostic process starts with the physician interviewing the patient (or his or her caretakers when the patient is unable to describe the complaints) to begin to consider what may be wrong. The clinician continues with the physical examination, gathering more information and all along listing the possible causes of the disorder. Next, laboratory tests will add evidence that moves the physician along the decision tree, strengthening some diagnostic possibilities and precluding others. Even after the treatment has started, the response to therapy may modify the rank that various hypotheses occupy.

Because the physician is following an orderly pattern, adopting and eliminating diagnostic hypotheses in accordance with the information that becomes available, in the twentieth century it was felt appropriate to design computer programs that could follow an algorithm; that is, could indicate which of two alternate directions to take depending on the existing data and thus reach a diagnosis. It was found that these programs could be valuable adjuncts in diagnosis but could not replace an experienced clinician. This is because a computer cannot detect the many nonverbal clues that may be garnered from a patient's history and physical examination, nor perceive the emotional tone that overlays the illness and that causes some symptoms to be emphasized and others to be minimized or obscured. In practice, the physician must sometimes take care first of those clinical problems that—he or she knows—pose a greater threat to life and health than those that happen to preoccupy the illness's sufferer most insistently.

It scarcely needs to be said that for the difficult diagnostic process to take place satisfactorily, a long and trusting relationship between patient and physician must exist. Unfortunately, in some industrialized countries, current social changes and the organization of health services at a national level have created conditions that erode this confidence.

8

THERAPY

Therapy (Greek *therapeia*, "attendance," or "medical treatment," from *therapeuein*, "to attend, to treat") is the entire set of actions and practices aiming to cure or palliate disease. Similarly, "therapeutic," in one definition, is "of or relating to the treatment of disease or disorders by remedial agents or measures." Clearly, the meaning of these terms is very general, since it covers "all measures, regardless of their nature, employed to minister to the sick." Included are such diverse curative methods as psychoanalysis, which uses words as tools to alleviate disturbances in the psychic life, and gene therapy, which consists of replacing a defective gene with a normal, functional gene, which is transported into the cell by various vectors (for instance, viruses) or forced into the cell by other means. Therefore, therapeutic measures include such a broad range of practices that a comprehensive description cannot be attempted here. In what follows, only a brief overview is presented of historically important therapies, some characterized by the use of active agents or medicaments and some that have attempted a different approach.

HERBAL MEDICINE

It is likely that plant medicaments are the most ancient form of therapy known. Two types of herbal medicine (as of medicine in general) seem to have existed for a long time. The first is popular, traditional, "folk" medicine, of strictly empirical origin and transmitted orally. The other is part of "official" medicine, developed by physicians who investigated the efficacy of the remedies, the reasons for their action, and their proper classification. "Folk" medicine is of extraordinary interest, but the historian finds it almost impossible to trace, due to the lack of reliable written sources. It should not be disparaged as irrational or superstitious. The fact is, considerable reasoning and observation went into its development—often as much, and as careful, as into "official" medicine. The only difference is that the concept of what constitutes disease varies with each. But historians and scientists have much to learn by turning their conscientious and deliberate attention to folk medicine.

The oldest herbal medicine is probably that of China. Shen Nong, a semilegendary personage who taught people how to identify and use plants for medicinal purposes, lived about 4,700 years ago. A medical treatise, the *Shen Nong ben cao jing,* bears his name, but it was written 2,500 years later. It lists 252 medicaments of vegetal origin, many of which are still in use today, and describes the characteristics and manner of use of these plants. The other ancient book of Chinese medicine, the "Yellow Emperor's Classic of Internal Medicine," or *Huang di nei jing,* also incorporates numerous herbal medicines.[1]

Folk medicine can be powerful and effective. For example, a plant known in China for two thousand years has recently been shown to be the most effective therapy against malaria. It is by no means an exotic plant and is known in the West as *Artemisia annua.* (The *Artemisia* family includes a species from which is distilled absinthe, brandylike spirits of high alcoholic content.) From decoctions of this plant Chinese researchers extracted the active product, which they named *qing hao su,* or "artemisinin." Until today, no treatment failures have been experienced with the use of artemisinin, but in 2006 the World Health Organization (WHO) warned pharmaceutical companies to refrain from issuing antimalarial products whose only active principle is artemisinin, in order to prevent the buildup of drug resistance by the parasite. This has happened with previous drugs, such as chloroquine, which was also 100 percent effective against malaria when introduced in Thailand in 1977, then lost much of its effectiveness.

About 40 percent of the world population is at risk of contracting malaria. Every year, 1 million to 3 million people die of this disease, and 300 million become infected. This should have been enough to mobilize the whole world against this scourge. Still, the response of the most developed societies has been muted. Sanitation measures and living conditions have virtually eliminated this disease from industrialized countries. It was the Vietnam War that, as a result of the many American casualties caused by this disease, demonstrated the low efficacy of conventional antimalarial agents and spurred American researchers in their quest for new ones. Meanwhile, the North Vietnamese turned for help to the Chinese, who, reviewing their dusty pharmacopoeia, came up with *qing hao su* in 1970. It had been forgotten by the world, perhaps due to the Western disdain for unofficial "folk" medicines and to the fact that artemisinin's virtues had originally been proclaimed in the midst of the Chinese Cultural Revolution, when it was wise to look upon Chinese proclamations with skepticism. Be that as it may, WHO declared in November 2001 that "the greatest world's hope to provide a treatment for malaria comes from China." And one of the officers in an antimalarial campaign stated, as reported in the popular press, that "given the present notoriety of artemisinin, it is unbelievable how much we ignored it [in years past]."[2]

In ancient times, Chinese knowledge of herbal medicine was passed on to Persia, Egypt, and India. In Greco-Roman antiquity, herbal medicine was also cultivated. In ancient Greece, a group of men, called "rhizotomists," were distinguished from physicians and were probably semishamanistic precursors of professional doctors. They were presumed to engage in black magic and to perform obscure rituals when gathering plants. Pedanius Dioscorides, a widely traveled army physician who lived in the first century A.D. under the emperors Claudius (10 B.C.–A.D. 54) and Nero (c. 41–c. 90), composed a book, *De materia medica,* that lists 1,000 medicaments, about 600 being plants. It became an authoritative medical text for several centuries, in the West as well as in the East.

On the American continent, the Aztecs possessed a vast knowledge of medicinal plants. As with the Greeks, there seems to have been a division between drug vendors, who brought their remedies to sell in the public market, and the educated class of physicians, who had received theoretical instruction on the properties and classes of remedies. This teaching usually passed from father to son and was heavily tinged with Aztec religious concepts. Among the Aztecs, mysticism was so inextricably entangled with all aspects of their lives that it is impossible to say where the sorcerer ended and the physician began.[3]

The Aztec emperor Motecuhzoma (Montezuma) I possessed extensive botanical gardens that were admired by the Spanish conquerors, who were quartered there during their campaign of invasion. These gardens were used for medical research, and the preparations made from the plants—juices, powders, extracts, infusions, syrups, plasters, and so on—were given free to patients, provided that they agreed to report the results. At least 1,200 different plants were used. Some were used as mixtures, thus increasing the number of preparations to incredible levels. However, some Spanish chroniclers criticized the Aztec physicians for their tendency to use single agents instead of the multidrug combinations exemplified by the famous "theriacs" (see below) common in Europe at the time. They also criticized Aztec medicine for its failure to practice bleeding, a procedure to whose "curative" virtues every self-respecting European physician of the sixteenth century was ready to swear.

The first text of herbal medicine produced in the American continent was written in Nahuatl, the language of the Aztecs, by Martín de la Cruz, a native of Tlatelolco, Mexico, who had been instructed in herbal medicine by the elders of his village. It was translated into Latin by Juan Badiano, also a Mexican native, was transported to Europe, and was there kept at the Royal Library in Madrid, under the name Badianus Manuscript. Historical vicissitudes determined that it should pass into the hands of Cardinal Francesco Barberini (1597–1679), who bequeathed it to the Vatican Library, where it was filed as Codex Barberini, Latin. Except by a few scholars, this manuscript remained ignored for nearly 350 years. Yet it is a document of extraordinary historical importance, since it reflects the medical tradition that existed in Mesoamerica before the arrival of the Europeans. In addition, 90 percent of the plants in the codex are still in use, especially in rural Mexico.

Based on the testimony of the friars who accompanied the Spanish forces in the conquest of Mexico, we know that the Aztec physicians could cure chronic diseases that resisted all the therapeutic efforts of European doctors.[4] There is evidence that the Aztecs had better treatments for the maladies that were common among them than the Spaniards did. They were experts at treating war wounds, in which they had considerable experience, since they lived in a state of perpetual warfare. Spanish military surgeons were still using the barbaric method of pouring boiling oil on or applying a red-hot iron to the wound. Ambroise Paré (see chapter 2) had not yet revolutionized the methods of most European armies for controlling bleeding. The Aztecs used extracts of a plant, *Commelina pallida,* that is now known to promote blood coagulation and vascular constriction. Interestingly, Chinese medicine also used *Commelina pallida* on wounds. That two unrelated cultures, very distant from each other, should have developed independently the same medical use for a plant lends credibility to the ancient documents and tells us that the remedy was truly effective.

They were also good at treating burns, some bacterial and fungal infections, and the bites of venomous animals, such as snakes, spiders, and scorpions. But an area of special attention for Aztec physicians was obstetrics. Plants that have since been scientifically proven to induce uterine contractions were used to hasten births. An example is *Montanoa tormentosa*, still in use in rural Mexico under the name of *cihuapatli*. This and other plants could also be used as abortifacients.

Bernard Ortiz de Montellano, author of the most comprehensive modern treatise on Aztec medicine, states that folk medicines are often given for the wrong reasons but are subsequently found to possess properties that make their use eminently reasonable. He cites the practice in some African tribes of giving freshly ejaculated semen to women in travail to drink, in order to hasten the childbirth. The basis of this primitive remedy is strictly magical, namely, the idea that "what caused the child to go in must be of help in letting him out."[5] However, it was subsequently found that semen is rich in prostaglandins, powerful inducers of uterine contractions, and the amount present in a single normal ejaculation is known to be effective, when taken orally, to induce labor at term.

The Aztecs used *Argemone grandiflora*, a plant similar to the opium plant, for the treatment of pain. Plants of the genus *Dioscorea*, now known to contain steroidal compounds, were illustrated in Codex Badianus and were used to treat inflammatory conditions. Curiously, the codex makes no mention of psychotropic drugs, in spite of the known use of hallucinogenic mushrooms by some native tribes of Mexico. It may be that the use of these plants was connected with rituals of the pre-Christian religions practiced in America before the European colonization and that prejudice against paganism, in the staunchly Catholic ambience of Spain's colonies, forced the compilers of the codex to exclude any mention of such plants.

Also from pre-Columbian America, although not from Mexico, was another vegetable product of considerable medical importance. The bark of the cinchona tree, originally from the Andean region of South America and also known as "Peruvian bark," "Jesuits' bark," or "quinquina," contained quinine and quinidine, drugs used in the treatment of malaria and disorders of cardiac rhythm, respectively, in addition to having some analgesic and antipyretic effects. Quinine, a white, bitter-tasting alkaloid, was the drug of choice for the treatment of malaria for a long time, until it was replaced by synthetic drugs such as quinacrine in the 1920s. It remains a useful drug when others prove ineffective. Quinidine, which is chemically related to quinine, also has some antimalarial potency but now is used primarily to combat heart rhythm disorders, as it slows down the electrical conductivity of the cardiac muscle.

There are many differing accounts of how the use of the Peruvian tree bark became widespread. Some say that the natives zealously guarded the remedy from their conquerors, but the secret came out when a compassionate Indian treated a wounded soldier who had sought refuge in his hut. Others claim that the first user outside the Indian community was a *corregidor* (provincial governor) named Juan López de Cañizares. It is generally accepted that years later, in the seventeenth century, when the count of Chinchón was the Spanish viceroy in Peru, either he or his wife fell ill with intermittent fevers, and the bark of the miraculous tree was the instrument of recovery. The important thing is that recognition of the therapeutic value of this tree was one of the achievements of early Native American herbal medicine. Surprisingly, this medicament was forgotten for many years in its native land and did not recover its popularity until it was reintroduced by physicians who had received their medical education in European universities.[6]

Demand for the product grew in Europe, where it was introduced by the Jesuits (hence the name "Jesuits' bark"). In England, a physician named Robert Talbor successfully treated King Charles II with this medicine, and thus gained great preferment at his court. Next, Talbor introduced the Peruvian bark into France, where it produced a veritable craze. Louis XIV was cured of a fever by the "English remedy" (*le remède anglais*). The poet La Fontaine wrote, in 1682, a "Poem to Quinquina," an unusual theme for a poetic composition, in which he refers to it as "a second Panacea" and a gift of Apollo to mankind.[7]

Another important plant remedy is mentioned by Pliny the Elder (23–79) in his *Natural History*. He talks about a dark plant "whose mere touch stanches a patient's bleeding; some call it *hippuris* and others *ephedron*." He adds, "Its property is to 'brace' the body [some translate "to make the flesh more compact"]. Its juice, kept in the nostrils, checks hemorrhage therefrom, and also checks looseness of the bowels. [It] ... promotes passing of urine and cures cough and orthopnea."[8]

Clearly, Pliny is alluding to the action of ephedrine, the compound isolated from plants of the genus *Ephedra*. Pliny described it as having no leaves but only long, segmented branches. Ephedrine was first isolated from plant extracts in 1887 by a Japanese investigator, Nagajosi Nagai, who reported a sympathomimetic action (i.e., similar to that of adrenaline) of this drug. However, a thorough investigation of the physiology, biochemistry, and medical uses of this plant product did not come until 1924. The Chinese researcher Ku-Kuei Chen and the young American physician Carl F. Schmidt, working at the Pekin Medical Union, reisolated the compound and thoroughly assessed its effects. They injected extracts of this plant into anesthetized dogs, which showed a marked rise in blood pressure. They determined that ephedrine is also a bronchodilator of help in treating asthma (at that time, only theophylline was known to have this effect), whooping cough, and emphysema. Its effects on blood pressure are useful in correcting hypotensive states. Unlike adrenaline, ephedrine could be administered orally, as described in ancient times by Pliny, who said it was given with wine.

The Chinese had been curing cough and stopping hemorrhages with the *Ephedra sinica* plant, which they called *ma huang,* for thousands of years. It is unlikely that the ancient Romans acquired this knowledge from the Chinese, of whom they knew next to nothing. This is another example of independent development in two cultures of medicinal uses for the same plant, since there is no evidence that the ancient Greeks or Romans had contact with the Chinese medical tradition.

Considering the millions of life forms that constitute the vegetable kingdom, including roughly 500,000 species of what we call "plants," there is a great potential for discovering new substances of medical usefulness. It is equally obvious that a summary of the work done in this area is a formidable task that cannot be attempted here. However, mention must be made of one more pharmacologic agent of plant origin that became central to modern medicine, namely, digitalis.

Recognition that a plant could cure "dropsy" (in the seventeenth and eighteenth centuries this designation encompassed bodily fluid accumulation due to cardiac, renal, or other origins) is generally attributed to the Scottish physician William Withering (1741–1799). In 1775, he had to treat a patient with difficulty breathing and swollen legs. Medicine could do nothing against this condition. But having heard of a folk remedy (from a gypsy, legend has it) that was effective in similar cases, he decided to try it and was rewarded by seeing that the patient eliminated much liquid and started to breathe more freely. The "secret recipe" contained many ingredients. However, Withering, who was well informed in botany, decided that foxglove (*Digitalis purpurea Linnaeus*), a plant common in Europe, had to contain the active principle. There are nineteen recognized species of foxglove. They are characterized by bell-shaped flowers, some quite elongated, like finger gloves.

Digitalis is cardiotonic: it reinforces the strength of contraction of the cardiac muscle and helps eliminate the body's excess fluid (edema). It is used in the treatment of cardiac failure and is effective against edema of cardiac origin; however, not when the water surfeit is due to renal insufficiency, as Richard Bright (1789–1858), a peerless clinical investigator of renal disease, would later prove. Withering administered to his patients infusions of the dried, powdered leaves, and his results were quite variable. The amount of digitalis varies with the plant species, with the season of the year, and even with different parts of the same plant. As he increased the dose, toxic levels were very promptly reached, and Withering had to learn to identify the signs of toxicity, such as vomiting, diarrhea, changes in vision, and mental confusion, so as to decrease the dose promptly when these manifestations appeared. The therapeutic and toxic levels were dangerously close to each other, and it was extremely difficult to determine the right dosage. This difficulty was lessened when synthetic, crystalline, perfectly standardized forms of the drug became available.

Antidotes: Theriacs, Bezoars, Mummies, and Other Historical Remedies

Seeing that plants and other natural products could be beneficial in combating illnesses, it was inevitable that people should try combinations of medicaments, in the hope of potentiating their curative effects. They devised remedies whose therapeutic virtues may be doubted but that attest to the robustness of the human imagination. The best example is what came to be called "theriacs," staples of pharmacology for several centuries: they could still be found in apothecary shops during the nineteenth century. The theriac (Greek *theriaké*, from *therion*, "savage beast," and *theriakos*, "good for the bites of savage beasts") originated in antiquity as a preparation against poisonous bites, such as those of spiders, scorpions, or rabid animals. In the time of the Roman Empire, when the military and bureaucrats moved from country to country, they may have resorted to theriacs as preventive measures, much as people today use vaccinations when traveling to endemic areas of infectious diseases.

There is a tradition that theriacs were started by King Mithridates VI Eupator (120–63 B.C.), ruler of Pontus, in northern Anatolia (now Ukraine), and the last ruler of that name. This man opposed the Roman might in Asia Minor but was vanquished by Pompeius. He must have been an impressive personage. A tall, strong, cruel satrap, his very name was calculated to impose: Mithridates means "gift of Mithras" (the most important Iranian god of the sun and deity of contracts and of war in pre-Zoroastrian times). Afraid of assassination by poisoning, he began drinking small doses of poison every day, so as to render it harmless; and he did so with a variety of poisons. According to Pliny, he was "the first to discover various antidotes, one of which even bears his name" (*"primo inventa genera antidoti ex quibus unum etiam nomen eius retinet"*).[9]

Indeed, the English language has retained the word "mithridate" or "mithridatum" to denote an antidote to a poison; and "mithridatize" means to become immune to a poison by ingesting gradually increasing doses of it. According to legend, when Mithridates was captured by Pompeius and ordered to take his own life by poisoning himself, he could not do so, for he had developed resistance to all manner of toxic products, and he had to be put to death by the sword.

The Romans, as is well known, made poisoning into a fine art. Emperors hired professional poisoners, as politicians today hire campaign managers, to take care of political opponents. But since they themselves were subject to this sort of expedited elimination, they created the role of "foretasters," or *praegustatores,* who assayed the toxicity of foods by ingesting a little and seeing what happened: a risky occupation with a fair demand in the job market. Foretasters became organized into guilds, with their own elected officials. Little wonder that the demand for new and better antidotes grew speedily in Roman times. But what started as poison antidotes soon blossomed into compounds allegedly useful in the treatment of the plague, malignant fevers, or other conditions. And pretty soon they were good for practically anything, including for the hale and strong, who took them preventively, as prophylaxis. Galen devoted a book to the subject, called *Theriaké.*

This is how the preparations known as "theriacs" entered into official medicine. The reasoning was that compounds expected to counteract an undesirable agent had to have at least some elements of the thing they were supposed to antagonize. Thus, for a theriac to work against snake bites, it had to contain snake skin, snake flesh, or some other reptilian ingredient. But imagination was always eager to add its own embellishments to medical therapy. Theriacs grew in complexity, to become cocktails of so many ingredients as to stagger the imagination. There could be between forty and seventy, including the most outlandish and heteroclite, such as viper's flesh, opoponax (an aromatic gum resin of a plant), beaver's kidneys, myrrh, licorice, oil of turpentine, saffron, and so on. Opium was commonly used in theriacs, perhaps because it was the only ingredient whose strong pharmacological action was sure to bring some manner of relief.

During the Middle Ages and the Renaissance, the exact composition of theriacs was zealously guarded. Each county, each city, had its own particular formula. Venice and Bologna were famous for the efficacy of their respective theriacs.[10] Secrecy augmented the esotericism that surrounded these medicaments and perhaps contributed to their placebo effect. Theriacs were highly regarded throughout the Western world; patients and physicians continued to look for them for centuries.

At the end of the Middle Ages, the Galenic teachings began to be questioned. Under the influence of the iatrochemical school of thought, the search for remedies was directed to the whole of Nature. No longer could the complex constitution of man be deemed explainable by the interplay of four humors. A consequence of this enlarged outlook was a diversification of the pharmacopoeia, with medicaments now being drawn from the animal, vegetable, and mineral kingdoms. But scientific reasoning had not yet imposed its much-needed orderly method, and the remedies continued to display a stunning bizarrerie and exuberant capriciousness.

Well known is the semimiraculous power that the Renaissance attributed to bezoars, calculous concretions formed from indigestible material, that can be found in the stomach of some animals, such as gazelles, goats, llamas, and other ruminants. Bezoars (from Arabic *bazahr*, Persian *padzahr*, "that which preserves from poison") were used in the treatment of melancholia, as well as against toxic states. According to the official inventory of the properties of Queen Elizabeth I of England, made upon the accession of her successor to the throne, she had "a great Bezoar stone set in goulde ... and some other large Bezoar broken in pieces." Bezoars were precious possessions, carefully stored, like jewels, and passed on from parents to children, like family heirlooms. They were still in use well into the eighteenth century.

Among exotic remedies used during the Renaissance were so-called Egyptian mummy and unicorn's horn. Of course, these were mostly counterfeit products. The former was supposed to be material found in ancient Egyptian tombs, either distilled from a mummy's wrappings or the mummified or skeletonized remnants themselves, appropriately ground and made into ingestible or topically applicable preparations. Ambroise Paré remarked that Egyptian mummy was sometimes made "in our France" and that such remedies "were as good as those brought from Egypt; because they are none of them of any value." As to the unicorn's horn, the fact that it was the appendage of a mythical animal that no one had ever seen did not preclude the high respect and extremely high price that a credulous age was willing to attach to so rare a commodity. Pope Clement VII (1478–1534), uncle of Catherine de Médicis, gave to her bridegroom's father, King François I of France, a piece of "unicorn's horn" (possibly part of a narwhal's tusk), reputedly a most powerful counterpoison.

ENEMAS AND THEIR HISTORICAL PROMINENCE

The routes of administration of medicaments were the same as today, except for the late development of intravenous drug delivery. The intrarectal medication is of considerable historical interest: it is surprising to discover the very large place that enemas or clysters occupy in the history of medicine. The enema (Greek ένεμα, Latin *infundo:* "to pour in") or clyster (Greek κλνστήρ: a clyster pipe or syringe, from κλυζ ειν: "to wash out or drench") is of ancient origin. It is said that primitive peoples used to wash the rectum using a hollow reed stem and the running water of a river current. The ancient Egyptians brought a high level of sophistication to this procedure. The Ebers Papyrus provides evidence for the use of enemas as far back as 1400 B.C. One legend, supported by such authorities as Pliny, Plutarch, and Galen, says that the Egyptians were taught this practice by the sacred ibis, the bird that was believed to use its long beak as a nozzle to self-administer an enema of sea water, as ancient authors put it, "whenever it feels incommodated by an excessive burden." Among ancient historians, Herodotus wrote of the Egyptians, "every month for three successive days they purge themselves, for their health's sake, with emetics and clysters, in the belief that all diseases come from what a man eats."[11]

An old engraving, purportedly showing the sacred Egyptian ibis self-administering an enema with its beak.

It is well known that the ancient Egyptians had medical specialists, if they may be so called in the prescientific era. According to Herodotus, "there were innumerable doctors, some specializing in the treatment of the eyes, others of the head, others of the teeth; and so on."[12] Less well known is that among the specialists, one was referred to as "guardian of the anus" (*nero pehut,* which some translate as "shepherd of the anus"), a title that authorized the physician to administer medicaments rectally.[13] The extraordinary importance that the ancient Egyptians attributed to colorectal diseases is not unique. Nor is the preoccupation with enemas restricted to this ancient culture. Malodorous, disgusting refuse is associated with decay: a dead body stinks as it decomposes; gangrenous, necrotic, and putrefying tissues are identified with all that is filthy and repugnant. It is not surprising that in the popular imagination death and refuse are indissolubly joined. Hence the compelling urgency to eliminate bodily waste if life and health are to be preserved. This idea has been more or less apparent throughout history. In the European seventeenth century, it was highly conspicuous. That era knew a veritable "clysteromania."

King Louis XIV is said to have received more than two thousand enemas, sometimes three to four a day, as meticulously recorded in the "Journal de la santé du roi." It became a routine, even a daily necessity.[14] And since the entire court wished to imitate the Sun King, the practice became widespread, first among the aristocracy, then among commoners. The duchess of Burgundy had a servant, Nanon Balbien, a former maid to Madame de Maintenon, give her an enema before attending a reception or a ball. According to the chronicler Saint-Simon (1675–1755; full name: Louis de Rouvroy, duke of Saint-Simon), this was done, at least once, even in the presence of the king, who could not see the proceedings taking place under the lady's ample skirt. However, observing that the maid was kneeling behind her, he became suspicious and asked what was going on. At last he was told. "What!" exclaimed the king. "You are taking an enema here!" "Why, certainly," she said. "How is that possible?" Upon this, they all burst forth laughing. The procedure was supposed to lighten ladies' complexions and to prevent headaches in the overheated palace halls where balls and receptions took place.[15]

Medical treatises of the seventeenth century demonstrate an amazing elaboration on this theme. We are told that clysters were classified according to their specific properties and composition. They could be emollient, purgative, astringent, anodyne, detersive, dividing, consolidative, or nutrient. Regnier de Graaf (1641–1673), the eminent Dutch physician and savant, whom we have previously mentioned as the discoverer of the egg cell in the human ovary (see chapter 4), wrote a whole treatise on enemas, *De clysteribus,* published in 1668. He had this to say about clysters of the "emollient kind":

> The emollient clysters, properly so called, are simple or composite. They are simple when they are made of a unique substance, such as milk, lukewarm water, or mutton broth. They are called composite when they are formed with various emollients, such as mallow, marshmallow, wall pellitory, violet, etc., with which one makes a decoction in ordinary water, milk, or any other emollient liquid. It is common to add to this decoction regular oil or oil of violet or sweet almonds, or else butter.[16]

This quotation gives an idea of the detail with which each type of clyster is described. Graaf also gives the specific medical indications for each kind and much additional information about this important remedy. However, he does not exhaust the modalities in use. Medicine at the time impresses as a patchwork of measures, elaborated locally by various individuals with little inclination to share their experience or results, and frequently contradictory. In England and various parts of Europe, tobacco-smoke enemas were advocated for cases of asphyxia by drowning, as well as for intestinal obstruction due to volvulus and incarcerated hernias. This practice, which may have started late in the sixteenth century, had not yet been abandoned by the second half of the nineteenth century.[17] A special apparatus was constructed for the purpose; it had a metal container in which the tobacco was burned, and the fumes were propelled by means of bellows, via flexible tubes. This peculiar procedure made the anonymous French translator of *De clysteribus* exclaim, in a footnote, "What a great pleasure for smokers, to be able to indulge by both ends!"

A Brief Note on Psychiatric Medicine and Its Therapies

Due to space constraints, psychiatry will receive only the most cursory mention in the present work. This discipline can hardly be considered a branch of medicine before the nineteenth century. However, the therapies used for psychiatric disorders are illustrative of the extraordinary evolution that the healing art has undergone since the inception of modernity. In ancient times, mental disease was often interpreted in a mythicoreligious context as possession by gods, demons, or other supernatural agents. Rather than a diagnosis of specialists, it was the society that decided who was mentally ill, and it was the family that took care of the afflicted. When restraint became necessary, the most violent and coercive means were applied; and when special housing for patients began to be available, the therapy remained primitive. The first lunatic asylums were gothic horrors where manacles, chains, or straitjackets were routine. In postrevolutionary France, Philippe Pinel (1745–1826) was a progressive reformist who experimented with softening or reducing the restraining measures, and this generated a more humane treatment of patients.

In Italy, this enforced gentleness was championed by Vincenzo Chiarugi (1759–1820), and in England by Samuel Tuke (1784–1857). The nineteenth century saw a rise in medical specializations and concurrently a greater sensitivity to the inhumane approaches of the past. By the middle of the nineteenth century, the English physician John Conolly (1794–1866) could publish his *Treatment of the Insane Without Mechanical Restraints*, with a favorable reception by the public. However, the treatment of mental patients in all the advanced countries continued to be a mixture of progressive measures, such as socialization and manual work, and ill-conceived, brutal procedures. These ran the gamut from cold showers and whirling chairs to electric shock, isolation, and dangerously copious bleeding.

In America, psychiatry followed a similar evolution. "Moral therapy" combined with various harsh measures became standard in a number of newly built institutions for the mentally ill. Among the most notable were the Hartford Retreat in Connecticut (founded in 1824), the MacClean Hospital in Boston (1818), and the Bloomingdale Asylum in New York (1821). In 1812, Benjamin Rush (1745–1813), widely considered the father of American medicine (he was, after all, one of the signatories of the Declaration of Independence), published a work entitled *Medical Inquiries and Observations upon the Diseases of the Mind*. Rush was a sharp clinician; his teaching of a thoughtful, deliberately conscientious, analytical approach to the patient remains to this day a mainstay of diagnosis. However, his recommendations for therapy, which included the restraining of mental patients, and the use of fear and extensive bleeding, have not resisted the test of time. The same may be said of Rush's advocacy of the ideas of John Brown (1735?–1788), the once highly influential but disquietingly strange medical theoretician (the late English historian Roy Porter has called him "the Scottish Paracelsus") who believed all diseases were but alterations of the "excitability" of the organism brought about by the conditions of the environment.

In France, under the recommendations of Pinel's followers, and among them especially Jean-Étienne-Dominique Esquirol (1772–1840), every *département* (administrative region of the French territory under a prefect) was required to have a public asylum for the treatment of the pauper insane. Esquirol believed that every mental disease was ultimately organic, that is, rooted in a structural or chemical abnormality of the brain ("somaticism"). But, as an heir to the clinical School of Paris, he performed extremely careful observations on patients, which led him to identify the social or psychological ("moral," as they were called then) triggers of mental breakdowns. These observations he recorded in a book, *Des maladies mentales* (1838).

In Germany, Wilhelm Griesinger (1817–1868) was also convinced that mental disorders were but "symptoms" or manifestations of diseases of the brain. However, he admitted a number of predisposing factors of psychiatric illness, including head trauma, heredity, febrile diseases, and "psychical causes." A respected historian, Erwin Ackerknecht, stated that with Griesinger "leadership in psychiatry passed into the hands of the Germans."[18] His followers adopted the somewhat dogmatic "somaticism" of their leader, although this one had at least attempted a synthesis of the physical and psychological causes of mental diseases.

Griesinger set psychiatry on a course whose explicit aim was to achieve a full scientific understanding of the organic basis of mental disease. Many of his followers conducted valuable research in neuroanatomy and neuropathology. The stress was upon the congruence of psychiatry and neurology. Distinguished physician investigators, such as Carl Wernicke (1848–1905), devoted their entire lives to this pursuit, and the importance of their contributions cannot be overestimated. To Wernicke are owed such insights as the cerebral localization of aphasias (conditions characterized by difficulty in the formulation or comprehension of speech), the dominance of one cerebral hemisphere or the other, and the correlation of neuropsychiatric manifestations with specific forms of brain damage.

Emil Kraepelin (1856–1926), elaborating upon the work of his predecessors, created a classification of mental diseases that Roy Porter called "the forerunner of today's *Diagnostic and Statistical Manuals*"[19] (the latter are the constantly updated volumes that list and define all the accepted mental diseases; today's "bible" for psychiatrists in all that pertains to nosology). This achievement was possible because Kraepelin adopted a "longitudinal" perspective of patients' histories; that is, instead of concentrating on the salient symptomatology (hallucinations, depression, and so on), he took into account the entire life history of the disease over time.

Unfortunately, all this exacting labor led nowhere in terms of treatment. A buoyant optimism had been generated in the late nineteenth century by the many brilliant investigations being conducted. But no effective cures were devised, not even a consistent amelioration—to say nothing of cure—of psychiatric illnesses. Declared one prominent clinician, German psychiatrist Georg Dobrick, with dismay, "We know a lot, and can do nothing."[20] Mental patients continued to languish in asylums, all but forgotten by the bulk of society and pining away in often dismal circumstances. It is not surprising that Nazi doctors, imbued with perverted ideas of eugenism and frustrated by the powerlessness of their best efforts to combat mental disease, came to believe that the deeply psychotic, like all patients with degenerative brain diseases, should be included among those who, they thought, were best eliminated.

From a more equanimous vantage point, the discouraging medical panorama also generated extreme measures, one of which was the development of "shock therapy." In the penetrating analysis of Roy Porter,[21] these therapies are marked by a disturbing ambivalence: on the one hand, they reflect the desperation of well-meaning psychiatrists to do something to improve the lot of forsaken, hopeless mental patients; on the other, they can easily turn these powerless sick persons into "experimental fodder" for reckless, insensitive doctors.

DESPERATE MEASURES:
SHOCK THERAPY AND PSYCHOSURGERY

Shock therapy presumably originated in antiquity. Convulsive phenomena are impressive manifestations that bring to mind seizure by gods, demons, or spirits. A patient who falls unconscious in an apparent state of trance, foaming at the mouth and shaking all over, intimidates the observer, suggesting the intervention of control by a supernatural power. The irrational belief that a convulsive patient is in the hands of a preternatural agency, joined with the observation that a sudden shock may bring an individual out of a state of lethargy and confusion, may have given rise to the idea of applying a strong stimulus as a way of mobilizing what has been called "the deepest sources of survival."[22] Some clinicians made reference to the observation—specious, as was ultimately shown—that patients afflicted with epilepsy rarely suffered from schizophrenia. There were even failed attempts to ameliorate the condition of the latter by giving them the serum of patients affected with a convulsive disorder. Camphor, an organic compound obtained from the wood of an East Asian tree, was also used as a medicinal. In large doses, it can induce convulsions. Its use as a convulsant is dangerous, however, since it requires toxic amounts too close to the lethal dose.

Electricity as a convulsive agent also has a relatively long history. It is said that the electric charge of a live torpedo fish (or "electric eel") was used in Greco-Roman antiquity for therapeutic ends, such as the blunting of the source of pain. According to the ancient narratives, the Roman emperor Claudius was so treated for headaches.[23] In the eighteenth century, Luigi Galvani (1738–1798) showed that sparking from an electrostatic generator could cause frog muscles to twitch, and Alessandro Volta (1745–1827) made similar experiments. Various monographs or articles subsequently appeared, authored by physicians who claimed to have discovered curative properties in electrical charges. William St. Claire (1752–1822) claimed to have cured "hysterical" convulsive seizures in a patient (this probably meant without demonstrable structural lesion; the term "hysterical" is virtually absent from present-day psychiatric nomenclature), and James C. Smyth (1741–1821), also using electricity, claimed to have successfully treated a patient with aphonia.

Many other physicians experimented with the medical uses of

electricity, but it was not until 1933 that Ugo Cerletti (1877–1963), a physician from Genoa, showed that an electrical current passed through the brain could be used to induce convulsions, as an alternative to insulin and Metrazol, at the time being given in convulsion-inducing doses for the treatment of schizophrenia. Electroconvulsive therapy (ECT) achieved better results, but it was a controversial procedure from the outset, as it causes memory disturbance and confusion. Experts in ECT have claimed up to 80 percent success in the treatment of depression. However, many find its drastic (some say brutal) action intuitively repellent. Furthermore, its mode of action remains unknown—some compare it to a broken-down car engine, where "slamming the hood makes it go."

In the United States, between 100,000 and 200,000 ECT treatments are given annually. From the 1940s to the 1960s, ECT was used extensively. Later it fell into some discredit, but in recent years it has shown a resurgence. This medical procedure received considerable publicity because of socially prominent patients receiving it: Vladimir Horowitz (1903–1989), the world-renowned pianist; James Forrestal (1892–1949), the first secretary of defense of the United States; Dick Cavett (b. 1936), the well-known television talk-show host; Zelda Fitzgerald (1900–1948), an artist and the wife of the author F. Scott Fitzgerald; and Vaslav Nijinsky (1888–1950), the famous ballet dancer. In some cases, the results were far from encouraging. The Nobel laureate author Ernest Hemingway (1899–1961), after being treated with ECT for two months at the Mayo Clinic in Rochester, Minnesota, committed suicide. "What is the sense of ruining my head and erasing my memory, which is my capital, and putting me out of business?" he asked, unhappy with the untoward effects of ECT. A similarly poignant description of the ECT experience was provided by the American poet Sylvia Plath (1932–1963) in her semi-autobiographical novel, *The Bell Jar*. She, too, committed suicide.

In an effort to make this therapy more humane, various technical improvements were introduced. The patient is first injected with a barbiturate and given a muscle relaxant to reduce the violence of the induced convulsions, which in the past were known to cause injuries such as bone fractures or collapsed vertebrae. On the other hand, these measures make it necessary to increase the intensity of the electric current, in order to achieve results comparable to those obtained before the administration of an anesthetic, and therefore the potential for untoward brain damage is increased. Inquests have shown that most patients treated with ECT report an improvement of their illnesses, but the value of this information is questioned on the grounds that it was collected from suggestible patients not long after undergoing ECT, who are still confused and who may not hold the same opinion later. Groups of concerned individuals have formed organizations that oppose ECT. In some countries, such as Japan, China, and the Netherlands, it is rarely employed. Italy, in 1999, passed legislation that severely restricts its use. However, the sheer reality is that there are profoundly psychotic patients who are resistant to virtually every other measure, including all available drugs, for whom ECT remains a last resort, the only one able to control a major psychotic episode.

However deleterious the untoward effects of this therapy may be, they probably lack the stark finality and irreversibility of brain surgery, which was also tried as a treatment for mental diseases. A number of observations, from disease-induced lesions to accidental injuries and animal experiments, suggested that interrupting the nerve fiber connections between the frontal lobes of the brain and the rest of the cerebral structure could cause alterations in mood and behavior. But it was António Egas Moniz (1874–1955), a Portuguese physician and neuroresearcher, who actually performed the first operation, on a female asylum patient in Lisbon. He called the operation a "leukotomy," since it severed the white substance; his American successors would later name it a "lobotomy," as it was supposed to cleave the frontal lobe, theoretically the site responsible for the behavioral derangement, from the rest of the cerebrum. Unfortunately, the recipients of these operations became passive, apathetic individuals who lost the ability to concentrate and showed little or no emotion. Egas Moniz was a cultivated man, a statesman in addition to his medical persona, who had been Portugal's ambassador to Spain and one of the signatories of the Treaty of Versailles upon the capitulation of Germany at the end of World War I. He received the Nobel Prize for Physiology or Medicine in 1949. Dr. Egas Moniz was shot in the back at sixty-five years of age by a schizophrenic patient of his but survived and died in his late eighties on the Portuguese farm where he was born.

The highly ambitious American physician Walter Freeman (1895–1972) avidly read the report of Egas Moniz's experience and took to building up his own case series with a fervor nothing short of evangelical. After initially collaborating with a neurosurgeon, James W. Watts, the partners had a falling-out. Having developed in cadavers a simplified procedure for the operation, Freeman decided to go it alone, despite his lack of surgical training. The procedure was as simple as it was horrible, grim, and revolting to watch. Under local or regional anesthesia, an instrument in every respect similar to an ice pick was hammered into the brain through the orbital roof, just above the eye globe. The ice pick, thus sunk about two inches into the brain substance, was manually waved in large arcs, so as to produce great swaths of devastation of the white substance. The patient was awake and was asked several questions, or directed to count backward; when the first indications of confusion appeared, the operation was stopped.

Freeman himself performed about three thousand lobotomies, traveling from state to state to give lectures and demonstrations. A number of his followers devised technical variants of the surgical operation. Tens of thousands were carried out throughout the world; between 1931 and 1951, 18,000 lobotomies were performed in the United States alone. Among the patients were famous persons. Frances Farmer, a film star and a beautiful and intelligent woman, was diagnosed with manic-depressive psychosis in a hospital where Walter Freeman performed lobotomies, which at the time were much favored as a fast and inexpensive means of controlling patients with unruly behavior; however, it is disputed whether she was one of the subjects operated on. In contrast, it is generally known that he did lobotomize Rosemary Kennedy (1918–2005), sister of the assassinated president John F. Kennedy (1917–1963). In a tragic episode, what is generally described as mild mental retardation associated with mood swings was thought to be correctable with lobotomy. The operation was performed, but the patient was left in a lamentable condition, much reduced in her mental capacity, incontinent, and incapable of living independently. She died at eighty-six years of age in an institution for the mentally handicapped.

Episodes of this nature generated widespread protest. What was once hailed by the media as a momentous medical breakthrough, a fast, clean way that "surgical wizardry" had discovered to restore sanity to the severely disturbed, was now exposed for what it really was: a reckless, irreversible mutilation inflicted on people and based on an overly simplistic conception of the workings of the human mind. By the mid-1950s, the operation fell into discredit and is currently condemned by the medical community.

PSYCHOANALYSIS AND RECENT PERSPECTIVES

There was another reaction, in the nineteenth century, against the hopelessness of existing therapies and the unyielding dogmatism of the "somaticists." This was the emergence of a more dynamic approach to mental disease, one that relied on psychological, not organic, explanations. It also made extensive use of hypnosis and thus harked back to the demonological or spiritual concept of mental disease. Eventually, it would develop into the doctrine of psychoanalysis. This is the only approach that, according to some scholars, permits us to look upon mental disease "in its primary intimacy, without necessarily placing it within the register of disease."[24]

Among its precursors, special mention must be made of Jean-Martin Charcot (1825–1893), who made extensive use of hypnotism and set up rather theatrical demonstrations at the huge Salpêtrière Hospital of Paris. Sigmund Freud (1856–1939) was then a young Austrian neurologist who came to study under Charcot. His time spent at the Salpêtrière would change his entire future orientation. Freud went back to Vienna, where in collaboration with Josef Breuer (1842–1925) he developed his theories on the sexual origins of neurosis and the disease then called "hysteria." In time, Breuer's ideas, with their emphasis on physiological concepts, departed from those of Freud, who stressed psychological mechanisms and the development of psychoanalysis.

Psychoanalysis looms large in the history of twentieth-century medicine. After Freud revealed the existence of a dynamic unconscious, repressed desires, the meaning of dreams, and infantile sexuality, the workings of the human mind could never be looked at in quite the same way as before. His theories spawned various currents of thought that posit a subliminal or potential madness in every individual. Psychoanalysts have proposed the concept that the child goes through, during normal psychological development, certain stages that resemble the pathological episodes of adults suffering from disturbed mental function, thus blurring the trenchant distinction between normal and pathological. Anthropologists have extrapolated this idea to types or stages of social organization in primitive societies, which presumably evolve through "oral" and "anal" phases. Conversely, ideas originating in anthropology have been used in attempts to illuminate the values and representations proper to children and to mental patients.

Thus, whatever limitations may be imputed to psychoanalysis—and it certainly has had its share of detractors—its creator, Sigmund Freud, remains one of the great intellectual figures of the twentieth century. His ideas exerted an influence beyond the confines of medicine and into fields as diverse as sociology, anthropology, philosophy, and cultural history. This influence was perpetuated in the work of such distinguished pupils as Alfred Adler (1870–1937) and Carl Gustav Jung (1875–1961), even when they discarded the ideas of the master to create their own systems.

More recently, the hopes of psychiatry have revolved around the development of mood-altering drugs. Starting with lithium, used since the 1940s against manic depression, a string of antidepressants and antipsychotic compounds has been made available. Valium, a tranquilizer, became the most widely prescribed drug in the world; Prozac, since its introduction in 1988, the most widely used antidepressant, with millions of people taking it around the world. The existence of potent psychopharmacological agents inspired the hope of doing away with the confinement of patients in asylums; treating the afflicted as ambulatory patients; markedly reducing the cost of their care; and ensuring their prompt reintegration into the community.

These hopes, unfortunately, have not been realized. Psychiatry remains a controversial field. In many parts of the world, the attitude toward the mentally ill remains one of exclusion and confinement. The latter tends toward concentration in large institutions and is often harshly coercive. Fundamental questions still await resolution. The debate continues on the definition of mental normality and how it differs from pathology. At the heart of this question lies the crucial problem of the causes and origins of mental illness: To what extent is it purely biological and modifiable by medical technology? Or is it an "epiphenomenon" issuing from social circumstances, especially the individual's family structure, education, and socioeconomic pressures?

Regardless of the answer, it is clear that industrialized society, which is the model imposed on most of the contemporary world, is rich in attributes that tend to destabilize the psychic function. Growing urbanization; fast-paced cultural changes that demand constant new acculturation; translocations of people, with loss of the support networks that once provided emotional protection; and the ever-present threat of addictions to various substances, some of which, like alcohol, are ubiquitous—all these factors portend the urgent need for some form of psychiatric medicine in the future.

9

SOME CONCLUDING
THOUGHTS

Those who take a panoramic ("Olympian," as some pretentiously put it) view of the history of medicine invariably discover a strikingly contrasting scene. From the ancient Greeks to the beginning of the twentieth century, the pattern is mostly one of desolation: in the past, at least one in ten babies died in infancy; rampant infectious diseases—tuberculosis, diphtheria, rheumatic fever, tetanus, pneumonia, meningitis, and other afflictions—mowed down lives in great swaths, without regard to age, sex, rank, money, or intelligence; women gave birth to their children among grievous pangs and ever-present dangers; and those coming into the world were not uncommonly destroyed before the first dawning of reason, for disease, infanticide, or death by neglect was the fate of numberless victims.

Medicine could do very little to alleviate these sufferings, beset as it was by almost absolute ignorance of the deadliest diseases and by the empirical nature of its treatments. Physicians could merely grapple with lethal ailments, console the afflicted, exhort survivors to resignation, and apply whatever fragmentary observations they could muster toward a meager alleviation. Up to the nineteenth century, there was but a handful of truly effective pharmacological agents: opium, quinine, digitalis, mercury (a poison as well as a treatment for syphilis and ringworm), colchicine (anti-gout), and not much else—not even aspirin, which was not synthesized until the end of that century and then mass-produced in the twentieth. "Of all the branches of human knowledge," wrote a Scottish historian in the second half of the nineteenth century, "medicine is that in which the accomplished results are most obviously imperfect and provisional."[1] He was undoubtedly right.

Since the physician was impotent in his appointed task, he was often mocked. His bleedings, fastings, clysters, and ventouses were the material of endless jokes, but at the same time, the people's expectations were low. They knew the doctor could do little; therefore, why blame him? Congruously, the physician's social stance was very limited; save notable exceptions, he (for the practitioner was almost always a male) did not wield much power. This was logical, since despite his best efforts his influence on the health of the population was minuscule. And one cannot help but wonder about the reasons behind the jokes played on him. For it is clear that mockery of the profession survives to the present, exerted by cartoonists, playwrights, and assorted satirists. Probably, the inability to stave off the diseases that torment us gave rise to symbolic rites of exorcism through laughter. To turn medicine and those who follow this calling into objects of derision has been, at bottom, a symbolic way to spoof the evil powers that menace us, or to keep them at a distance by dint of gags and trifles.

But the scene changed most dramatically within the last hundred years, when medicine's impact on society has been profound. At a time when the world's population was roughly doubling every forty years, medical research produced contraceptive pills that not only curbed demographic expansion in industrialized countries but affected the position of women in society and radically altered traditional sexual mores. On various fronts, medical progress challenged old assumptions and age-entrenched attitudes. Received ideas were suddenly obsolete. Reproductive technologies forced a reconsideration of the most cherished concepts of family, maternity, and filiation. The discovery of penicillin alone may be accredited with the survival of millions of individuals who

In the old tradition of satirizing the medical profession, this nineteenth-century caricature by Honoré Daumier (1808–1879) reads, in a bastard Latin taken from Molière's play The Imaginary Patient: *"First to bleed, then to purge, and later [or 'posteriorly'] a clyster to give."* COURTESY OF THE NATIONAL LIBRARY OF MEDICINE

might have succumbed without it; and a number of antimicrobials, hormones, and other potent pharmacological agents have followed that have altered the existence of millions. Medicine could thus change society and curb or enhance population growth, according to our own deliberate intent. According to some observers, the medical breakthroughs of the last fifty years have been responsible for saving more lives than during any other epoch since the beginning of medicine.[2]

Clearly, medical science has also made it possible for us to live longer lives now than at any other time in history. In the United States, life expectancy rose from 62.9 years in 1940 to 76.7 years in 1990; although this increase reflected in part the lowered infantile and childhood mortality rates (the result of simple measures of improved nutrition and sanitation), the gains at the upper end of the age scale were no less impressive. For instance, at age 65, life expectancy was 13.9 years in 1950 but rose to 17.8 years in 1998. Those who had reached the age of 75 in 1998, given their evident above-average vitality, could expect to live, on average, for 13 more years.[3]

The recent advances caused a widespread, unrestrained optimism. It is not difficult to find claims that formidable, hitherto insurmountable medical problems are going to be overcome in a matter of years or decades. These include genetic diseases (since gene modification techniques are already available), malignant tumors, and severe neurological damage, such as that sustained in paralysis-producing traumas. The media contribute to drawing the image of this utopia when they salute every new technological improvement as a "scientific breakthrough," often in terms as praiseful as they are nescient.

The result of this is an extraordinary demand for medical services that has expanded the medical establishment beyond the wildest anticipations. The pressures on the profession are enormous. People have been led to believe that modern medicine makes possible a paradisiacal existence, in which everyone has an imprescriptible right to physical beauty ("designer bodies" through plastic surgery), vigor, and painless, decay-free longevity. The most easily suggestible and ill-informed even talk in earnest of immortality as an attainable goal. Thus physicians have, in a sense, fallen victims to their own success and must daily confront demands for treatment of every conceivable complaint, including those related to normal physiological processes (for instance, menopause); every life event appears to have been "medicalized."

However, various alarming voices persist in sending us disquieting messages that dim this shiny picture. Iatrogenic diseases have grown disproportionately. New epidemic diseases, for which we are sorely unprepared, such as those caused by the Marburg and Ebola viruses, or the resurgence of old ones, such as influenza, threaten our lives. Afflictions of the elderly, such as Alzheimer's disease, are far from being understood, let alone controlled. Indeed, chronic diseases have replaced acute infectious diseases as leading causes of death. The most prevalent bodily disorders are now crippling ailments that target primarily older adults, such as cardiac disease, which in 1998 accounted for 31 percent of all deaths in the United States; malignant tumors (23.2 percent); and cerebrovascular accidents (6.8 percent).[4] A pyrrhic victory it would be, for medicine, if it had prolonged our lives only to expose us to new forms of misery. Thus euphoria turns to anxiety when it is realized that, in some cases, medicine has only bought us more time to be ill.

To these problems is added the influence of extraneous pressures that further distort the image of medicine and blur the nature of its role in society. Where does the function of medicine end? Spurious interests appear to try to "medicalize" every aspect of human life. The media, the relentless advertising of powerful pharmaceutical companies, and not least the medical profession itself exert a formidable influence to expand the diagnosis of treatable diseases, in such a way as to encompass the highest possible number of potential patients. Recently *The New York Times* published an editorial critical of the efforts of the American Society of Hypertension to broaden the definition of hypertensive cardiovascular disease. The proposed concept would define this disorder based not purely on the figures recorded in blood pressure readings, as was already being done, but also on other criteria, such as certain risk factors and biochemical markers.[5] Millions of persons who are presently not classified as hypertensive would fall into this category, according to the new criteria.

The editorial criticism found justification in the fact that the efforts to frame a new definition were heavily financed by major pharmaceutical companies. These firms are clearly interested in selling more drugs to more people. Although the American Society of Hypertension did not make a recommendation for a specific treatment, it seems obvious that a conflict of interest arises whenever drug companies are allowed to fund—at most generous levels—activities such as dinner lectures geared to promote the new definition. Developments of this kind lend credibility to the accusation that the medical-industrial establishment, "medibusiness," and the mass media are set on an unrestrained course to medicalize everything, under the premise that *everyone* has *something* wrong with them—something, of course, that must be treated.

Today, as before, the demand for medical aid that is addressed to a physician, and more generally to any healer, is deeply rooted in trust. For a treatment to work optimally, the sufferer must have faith in its efficacy and the competence of the therapist. That there is a large element of faith and suggestion here is undeniable; and that this element is helpful is a tenet universally acknowledged. This needs no reiteration; the impressive reality of the placebo effect constitutes the best proof of the truth in this assertion. But for trust and belief to grow, a strong interpersonal relationship must first be established, and this in turn requires the much-talked-about dialogue between patient and physician, which is being thwarted, it seems, by prevailing conditions in the organization of health care. At least in some Western countries, the industrial model that has been imposed has radically altered the traditional doctor-patient relationship.

Under this model, the doctor becomes a "health care provider" and the patient is a "customer." Since the middle of the twentieth century, certain schools of thought in sociology have accused organized medicine of professional dominance and of stigmatizing and victimizing defenseless members of the population who ask for its help. Arguing that the old terms "doctor" and "patient" carried negative connotations of dominance, the vocabulary was changed. Correlative to the language change, the methods and systems of industry were applied to medicine for the sake of enhanced efficiency. But this extrapolation has often been uncritical, so that under the new model the approach to medicine has become that of industry. This is counterproductive. Where the industrial model is most thoroughly implanted, the result is that all medical acts are rigorously codified and the doctor-patient relationship is reduced to a minimum. In some settings, the idea is to see the greatest number of patients in the shortest possible time, ominously evoking images of the assembly line. The psychological soothing skills and the pastoral forms of consolation that physicians of yore sedulously cultivated and were prodigal in dispensing to their patients (for lack of any more effective curative means) would now be ranked as "nonreimbursable" medical activities.

The sick person, however, sees his or her illness from a perspective that is alien to these perplexities. For the person who suffers it, disease is an experience that falls outside every rational or scientific consideration. The sufferer will address his most earnest pleas to any agent deemed capable of removing him, or her, from the painful, anguishing situation in which he or she is mired. Today, as before, the sick person starts by calling the physician. But in this era of inflated medical expectations, the image of the physician is changed and may not answer to the sufferer's deepest yearnings. Hence the appeals to alternative forms of medicine, which have existed from time immemorial along with the more "orthodox" or "mainstream" varieties of the healing art. They coexist today with the reputedly more "scientific" medicine, and, let us remark by the way, have historically done so in perfect mutual intelligence.

Scholars have noted that epochs characterized by an extraordinary development of science and technology are also the times most propitious to the proliferation of irrational popular manifestations and a surge of interest in the occult, mysterious, and unexplainable. Mircea Eliade noted that the last quarter of the twentieth century, a time of outstanding achievement in space technology, biological science, and other conquests of the rational mind, coincided with an explosive resurgence of popular interest in occultism, divination, mythologies, and esoteric doctrines.[6] This situation continues today. Simultaneously with the exploration of outer space and the performance of formerly undreamed-of organ transplants, an unprecedented number of people make a living reading tarot cards or practicing astrology. The number of horoscopes read in television programs or published in printed form, and of people consulting them to guide their conduct, is today greater than at any time in the past.

Medicine could not remain a stranger to this phenomenon. Crushed by a burden greater than they can bear and unable to find in "scientific" medicine the spiritual solace and attention to their needs, patients turn to alternative therapies that offer a holistic care. Hence the unparalleled popularity of Oriental medicine, yoga, meditation, homeopathy, and all the treatments that impress patients because of their stamp of originality and esotericism. Some are new, but some are old, reborn under a new guise. The history of medicine brings to light the "circularity" of many of these practices. Acupuncture, for instance, is often thought of as a new importation from the Orient, but it was brought to the West by the Jesuit missionaries who visited the

court of China in the seventeenth century and by the Dutch navigators who established commerce with that nation. There is evidence
that detailed descriptions of acupuncture and moxibustion have existed in the West for a long time,[7] subject to alternating periods of
popularity and obscurity.

Medicine faces external and internal challenges. The former are
technological and scientific problems; the latter, the quandaries that its
own structure and development have engendered. The first ones are
formidable, since they include such profound enigmas as the function-
ing of the mind and the finest details of our genetic persona. But if past
performance has any predictive value, we may trust human inventiveness and ingenuity, through whose awesome power the toughest difficulties have been resolved. As to the second type of problem, those
that refer to the core of medicine itself, much introspection will be required. Physicians will have to define the limits of their own field.
They must provide clear answers as to what constitutes disease (in the
case of psychiatry the question is far from idle) and how far their efforts should extend. They must come to grips with age-old existential
questions and decide whether a world without disease is possible or
even desirable, or whether death and disease are inevitable and the
doctor's role is merely to minimize suffering and make our burden
more bearable and more humane. In any case, they must find ways of
reinforcing the skills and attitudes that make a doctor caring and understanding of the needs of the whole human being. Whether such
skills can be taught, and this in addition to the heavy load of facts required to practice medicine, is not the least of the problems that the
profession faces, especially in a system of medical care that seems ad-
verse to the fostering of these very qualities.

ACKNOWLEDGMENTS

I am grateful to Will Murphy, whose many suggestions and careful editing enhanced the clarity and readability of this book.

I owe many thanks to my wife, Dr. Wei Hsueh, for her patience and unflagging support during the long hours of the writing of this work.

NOTES

1. THE RISE OF ANATOMY

1. Maurice Leenhardt, *De Komo* (Paris: Gallimard, 1947), pp. 54–70.
2. For the life and work of Herophilus, see Feridum Acar et al., "Herophilus of Chalcedon: A Pioneer in Neuroscience," *Neurosurgery* 56, no. 4 (April 2005): 861–867. For the famous library of Alexandria, see Paul H. Chapman, "The Alexandrian Library: Crucible of a Renaissance," *Neurosurgery* 49, no. 1 (July 2001): 1–14. Especially valuable for Erasistratus's insights into cardiac valves and cardiac anatomy is R.C.S. Harris, *The Heart and the Vascular System in Ancient Greek Medicine, from Alcmaeon to Galen* (Oxford, England: Clarendon, 1973).
3. Paul Moraux, *Galien de Pergame: Souvenirs d'un médecin* (Paris: Les Belles Lettres, 1985), pp. 113–114.
4. For an overview of Islamic medicine, see Lucien Leclerc, *Histoire de la Médecine Arabe* (2 vols.) (Paris: Ernest Ledoux, 1876); Manfred Ullman, *Islamic Medicine* (Edinburgh: Edinburgh University Press, 1997); John R. Hays, *The Genius of Arab Civilization: Source of Renaissance* (New York: New York University Press, 1975) (also the edition by MIT Press, 1983); Edward Granville Browne, *Arabian Medicine* (Cambridge, England: Cambridge University Press, 1962).
5. Colin A. Ronan (abridger-editor), *The Shorter Science and Civilisation in China,* vol. 1: *An Abridgement of Joseph Needham's Original Text* (Cambridge, England: Cambridge University Press, 1st paperback ed., 1980), pp. 71–73.
6. Nancy G. Siraisi, *Medieval and Early Renaissance Medicine: An Introduction to Knowledge and Practice* (Chicago: University of Chicago Press, 1990) (hereafter referred to as *Medieval and Early Renaissance Medicine*).
7. Charles D. O'Malley, *Andreas Vesalius of Brussels* (Berkeley: University of California Press, 1964). This work, by a distinguished professor of medical history at the University of California, although published more than

forty years ago, probably remains the best English-language biography of Vesalius.

8. Meyer Friedman and Gerald W. Friedland, *Medicine's 10 Greatest Discoveries* (New Haven, Conn.: Yale University Press, 1998).

9. Roger Caillois, *Au cœur du fantastique* (Paris: Gallimard, 1965).

10. An English translation of this chapter is available; see R. J. Moes and Ch. D. O'Malley, " 'On Those Things Rarely Found in Anatomy,' an annotated translation from the *De re anatomica* (1559)," *Bulletin of the History of Medicine* 34 (1960): 508–528.

11. Leonard D. Rosenman, "Facts and Fiction: The Death of Saint Ignatius of Loyola," *Surgery* 119, no. 1 (1996): 56–60.

12. Antonio Mezzogiorno and V. Mezzogiorno, "Bartolomeo Eustachio: A Pioneer in Morphologic Studies of the Kidney," *American Journal of Nephrology* 19 (1999): 193–198.

13. "Fabricius ab Aquapendente (1537–1619), Preceptor of Harvey," *The Journal of the American Medical Association* 198 (October 1966): 178–179. See also A. H. Sculetus, J. L. Villavicencio, and N. M. Rich, "Facts and Fictions Surrounding the Discovery of Venous Valves," *Journal of Vascular Surgery* 33, no. 2 (February 2001): 435–441. A comment on this article appeared in *Journal of Vascular Surgery* 33, no. 6 (June 2001): 1317.

14. Fabricius of Aquapendente, *The Embryological Treatises of Hieronymus Fabricius Aquapendente*, with introduction, translation, and commentary by Howard B. Adelman (2 vols.) (Ithaca, New York: Cornell University Press, 1942, reissued 1967).

2. The Rise of Surgery

1. For the history and bibliography of trepanation in ancient Peru, see Raúl Marino, Jr., and Marco González-Portillo, "Preconquest Peruvian Neurosurgeons: A Study of Inca and Pre-Columbian Trephination and the Art of Medicine in Ancient Peru," *Neurosurgery* 47, no. 4 (October 2000): 940–950. A helpful and scholarly discussion of prehistoric diseases and ancient medicine is in Don Brothwell and A.T.D. Sandison, *Diseases in Antiquity* (Springfield, Ill.: Charles C. Thomas, 1967).

2. Quoted in Marino and González-Portillo (ibid.): F. Graña, E. B. Rocca, and L. R. Graña, *Las trepanaciones craneanas en el Perú en la Epoca Prehispánica* (Lima, Peru: Imprenta Santa María, 1954); S. A. Quevedo, "Un caso de trepanación craneana en vivo, realizado con instrumentos precolombinos del Museo Arqueológico," *Revista del Museo del Instituto de Arqueología* (Peru) 22 (1970): 1–73.

3. Guido Majno, *The Healing Hand: Man and Wound in the Ancient World* (Cambridge, Mass.: Harvard University Press, 1975) (henceforth referred to as *Healing Hand*). This marvelous, highly engrossing study of the medical and surgical concepts of the wound in antiquity is a classic that will remain indispensable to students of ancient medicine for many years to come. Not content with referring to the written sources, Dr. Majno, a distinguished pathologist, submits the procedures described in the ancient texts to the tests available in a modern laboratory. Thus, he actually tests and illustrates the usefulness of the heads of certain ants (*Eciton burchelli* and others) as surgical staples. See also E. W. Gudger, "Stitching Wounds with the Mandibles of Ants and Beetles: A Minor Contribution to the History of Surgery," *The Journal of the American Medical Association* 84 (1925): 1861–1864.

4. Of the seventy volumes of the *Collectiones medicae* (Medical Compilations), only twenty-five survive. An English translation from the Greek of two of them recently appeared: Mark Grant, *Dieting for an Emperor: A Translation of Books 1 and 4 of Oribasius Medical Compilations with an Introduction and Commentary*, vol. 15, *Studies in Ancient Medicine* (Leiden: Academic Publishers E. J. Brill, 1997).

5. For the contributions of Paulus Aegineta to surgery, with especial emphasis on plastic surgery, see Raffi Gurunluoglu and Aslin Gurunluoglu, "Paulus Aegineta, A Seventh Century Encyclopedist and Surgeon: His Role in the History of Plastic Surgery," *Plastic and Reconstructive Surgery* 108, no. 7 (December 2001): 2072–2079. Reference to the English translation of Paulus Aegineta's work is included in the bibliography of this article.

6. M. Pasca, "The Salerno School of Medicine," *American Journal of Nephrology* 14, nos. 4–6 (1994): 478–482; E. de Divitiis, P. Cappabianca, and O. de Divitiis, "The 'Schola Medica Salernitana,' the Forerunner of the Modern University Medical Schools," *Neurosurgery* 55, no. 4 (2004): 722–744 (discussion on pp. 744–745).

7. Siraisi, *Medieval and Early Renaissance Medicine*, p. 161.

8. "Vie d'Ambroise Paré," chap. 2 in *Ambroise Paré: Textes choisis. Presentés et commentés par Louis Delaruelle & Marcel Sendrail* (Paris: Les Belles Lettres, 1953), p. 27.

9. For a well-crafted recent biography of John Hunter, see Wendy Moore, *The Knife Man: The Extraordinary Life and Times of John Hunter, Father of Modern Surgery* (New York: Broadway Books, 2005). Older works include J. Dobson, *John Hunter* (London: E. S. Livingstone, 1969); Gloyne Roodhouse, *John Hunter* (London: E. S. Livingstone, 1950); and J. Kobler, *The Reluctant Surgeon* (London: W. Heinemann, 1960).

10. Majno, *The Healing Hand;* see his note 81 to chapter 6, page 508.

11. F. S. Haddad, "The *spongia somnifera*," *Middle East Journal of Anesthesiology*
17, no. 3 (2003): 321–327. See also M. S. Takrouri and M. A. Seraj,
"Middle Eastern History of Anesthesia" (editorial, historical article), ibid.
14, no. 1 (1997): 3–6; Takrouri and Seraj, "Middle Eastern History of
Anesthesia," ibid. 15, no. 4 (2000): 397–413. A viewpoint expressing skepticism about the effectiveness of the inhalation methods of the medieval
Arab physicians is P. Prioreschi, "Medieval Anesthesia. The *spongia somnifera*," *Medical Hypotheses* 61, no. 2 (2003): 213–219.

12. F. Darwin (ed.), *Charles Darwin: His Life Told in an Autobiographical Chapter
and in a Selected Series of His Published Letters* (2 vols.) (New York:
D. Appleton & Co., 1898), pp. 11–12. Also published as *The Life and Letters
of Charles Darwin* (2 vols.) (New York: Basic Books, 1959).

13. William Bullein, physician and surgeon, in 1562. Quoted in Christopher
Lawrence, "Medical Minds, Surgical Bodies: Corporeality and the
Doctors," chap. 5 in Christopher Lawrence and Steven Shapin, *Science
Incarnate: Historical Embodiments of Natural Knowledge* (Chicago: University
of Chicago Press, 1998), p. 183.

14. Matthew Turner, *An Account of the Extraordinary Medicinal Fluid, Called
Aether* (London: J. Wilkie, undated edition, c. 1761); may be read online in
a Project Gutenberg e-text (e-book no. 12522) at www.gutenberg.org./
dirs/etext93/pimil10.txt (visited February 3, 2006).

15. A biographical article on William T. G. Morton, gathered from recollec-
tions of his relatives, is E. L. Snell, "Morton's Discovery of Anesthesia,"
The Century: A Popular Quarterly 48, no. 4 (August 1849): 584–589. May
be read online at www.cdl.library.cornell.edu/gifcache/moa/cent/
cent004800594.TIF6.gif (visited October 11, 2005).

16. H. J. Bigelow, "Insensibility During Surgical Operations Produced by
Inhalation," *Boston Medical and Surgical Journal* 35 (1846): 379–382.

17. James Young Simpson obituary, *British Medical Journal*, May 14, 1870,
p. 505.

18. Charles D. Meigs, *Females and Their Diseases* (Philadelphia: Lea &
Blanchard, 1848), p. 40.

19. For a review of the theme of "floating kidneys," see Sandra Moss,
"Floating Kidneys: A Century of Nephroptosis and Nephropexy," *The
Journal of Urology* 158, no. 3 (September 1997): 699–702. See also D. L.
McWhinnie and D. N. Hamilton, "The Rise and Fall of Surgery for the
'Floating' Kidney," *British Medical Journal* 288 (1984): 845. For a report of
recent experience with this disease, see Osama M. Elashry, Steven Y.
Nakada, Elspeth M. McDougall, and Ralph C. Clayman, "Laparoscopic
Nephropexy: Washington University Experience," *The Journal of Urology*
154 (November 1995): 1655–1659.

20. Kathleen Rice Simpson and Kathleen E. Thorman, "Obstetric 'Conveniences': Elective Induction of Labor, Caesarean Birth on Demand, and Other Potentially Unnecessary Interventions," *The Journal of Perinatal and Neonatal Nursing* 19, no. 2 (April–June 2005): 134–144.

21. Ibid. See also American College of Obstetricians and Gynecologists, *Surgery and Patients' Choice: The Ethics of Decision Making,* Committee Opinion No. 289 (Washington, D.C.: ACOG, 2005); W. F. Rayburn and J. Zhang, "Rising Rates of Labor Induction: Present Concerns and Future Strategies," *Obstetrics and Gynecology* 100 (2002): 164–167.

22. Roberto Heros and Jacques Morcos, "Cerebrovascular Surgery: Past, Present, and Future," *Neurosurgery* 47, no. 5 (November 2000): 1007–1033.

23. Denis Kalette, "Boy Going Home After Six-Organ Transplantation," *The Washington Post,* September 2, 2005.

24. C. S. Lewis, *The Problem of Pain* (New York: Touchstone Books, Simon & Schuster, 1996), p. 14.

25. Donald Caton, *What a Blessing She Had Chloroform: The Medical and Social Response to the Pain of Childbirth from 1800 to the Present* (New Haven, Conn.: Yale University Press, 1999). This work is especially valuable for the breadth of its approach to the discovery of anesthesia, the clarity and scholarliness of its presentation, and as a source of bibliography for the historical period discussed.

26. John Snow, "On the Administration of Chloroform During Parturition," *London Association Medical Journal* (1853): 100; Snow, "On Asphyxia and on Resuscitation of Stillborn Children," *London Medical Gazette* (1841–42): 222–227.

27. S. H. Calmes, "Virginia Apgar: A Woman Physician's Career in a Developing Specialty," *Journal of the American Medical Woman's Association* 39 (1984): 184–188. For the "Apgar score," see Virginia Apgar, "Proposal for a New Method of Evaluation of Newborn Infants," *Anesthesia and Analgesia* 32 (1953): 260–267.

28. J. A. Martin, B. E. Hamilton, P. D. Sutton, S. J. Ventura, F. Menacker, and M. L. Munson, "Births: Final Data for 2002," *National Vital Statistics Report* 52 (2003): 1–113.

29. Michael F. Greene, "Vaginal Birth After Cesarean Revisited" (Editorial), *The New England Journal of Medicine* 351, no. 25 (December 16, 2004): 2647–2649. See also the article by Mark B. Landon and twenty collaborators: "Maternal and Perinatal Outcomes Associated with a Trial of Labor After Prior Cesarean Section," ibid., pp. 2581–2589.

30. For the out-of-phase reaction of the public to progress in anesthesia, see especially the thoughtful discussion in Caton, *What a Blessing She Had Chloroform,* pp. 200–233.

3. VITALISM AND MECHANISM

1. Biographical data on Stahl, containing primary and secondary bibliographic sources, may be found in Lester S. King, "Stahl, Georg Ernst," *Dictionary of Scientific Biography*, ed. C. C. Gillespie (12 vols.) (New York: Charles Scribner's Sons, 1970–1976), vol. 12, pp. 599–606.

2. A discussion of Stahl's physiological ideas may be found in Lelland J. Rather, "G. E. Stahl's Psychological Physiology," *Bulletin of the History of Medicine* 53 (1961): 37–49. See also, for a larger context, Roger K. French, *Robert Whytt, the Soul and Medicine* (London: Wellcome Institute for the History of Medicine, 1969), pp. 117–148. The collected works of Stahl may be consulted in French translation in Georg Ernst Stahl, "Vraie théorie médicale," in *Oeuvres médico-philosophiques et pratiques* (4 vols., translated and commentary by Théodore Blondin) (Paris: J. B. Baillière et Fils). See vol. 3, 1860; see also vol. 2, Georg Ernst Stahl, "Mixte et vivante," ibid., pp. 366–376.

3. Extracts of Stahl's works *Zymotechnia fundamentalis seu fermentationis theoria generalis* (1697); *Schriften von der Natur des Salpeters* (1734); and *Experientia, observationes, animadversiones chimicae et physicae* (1731) were published in Henry Marshal Leicester and Herbert S. Kickstein, *A Sourcebook in Chemistry, 1400–1900* (New York: McGraw-Hill, 1952).

4. The concept of "tonic motion," according to Stahl, was the subject of an extensive study by Ku Ming (Kevin) Chang: "*Motus tonicus*: Georg Ernst Stahl's Formulation of Tonic Motion and Early Medical Thought," *Bulletin of the History of Medicine* 78, no. 4 (Winter 2004): 767–803.

5. Louis Dulieu, "François Boissier de Sauvages (1706–1767)," *Revue d'Histoire des Sciences et Leurs Applications* 22 (October–December 1969): 303–322. See also Lester S. King, "Boissier de Sauvages and 18th Century Nosology," *Bulletin of the History of Medicine* 40 (1966): 43–51.

6. Elizabeth L. Haigh, "Vitalism, the Soul, and Sensibility: The Physiology of Théophile Bordeu," *Journal of the History of Medicine* 31 (1976): 30–41.

7. Elizabeth L. Haigh, "The Vital Principle of Paul-Joseph Barthez: The Clash Between Monism and Dualism," *Medical History* 21 (1977): 1–14.

8. Examples of this author's obscurity, even in translation, include Jan Baptiste Van Helmont, "The Chief or Master-Workman," in *Oriatrike or Physic Refined*, trans. J. C. Sometime (London: Lodovick Loyd, 1662), pp. 35–36; and "The Seat of the Soul," in ibid., pp. 192–197.

9. Walter Pagel, "Harvey and Glisson on Irritability with a note on Van Helmont," *Bulletin of the History of Medicine* 41 (1967): 497–514. See also Owsei Temkin, "The Classical Roots of Glisson's Doctrine of Irritation," in ibid. 38 (1964): 297–328.

10. Albrecht von Haller's important work in Latin, *Prima lineae physiologiae*,

was translated into English as *First Lines of Physiology* by William Cullen (New York: Johnson Reprint Co., 1966) (two vols. in one, reprint of the 1786 edition). His work entitled "Sermones de partibus corporis humani sentientibus et irritabilibus" (Göttingen, 1753) was translated as "A Dissertation on the Sensible and Irritable Parts of Animals," by Owsei Temkin, *Bulletin of the History of Medicine* 4 (1936): 651–699.

11. Von Haller's collected investigations on sensibility and irritability exist in French translation by Ch. Tissot: Albrecht von Haller, *Mémoires sur la nature sensible et irritable des parties du corps animal* (4 vols.) (Lausanne: F. Grasset, 1762).

12. This definition was set down by Bichat in his *Recherches physiologiques sur la vie et sur la mort*, published in 1880. The first and major part of this treatise, together with Bichat's other writings on anatomy and physiology, were reprinted by G. F. Flammarion, Paris, in 1994.

13. "Treatise of Man," in René Descartes, *The World and Other Writings*, Cambridge Texts in the History of Philosophy, trans. and ed. by Stephen Gaukroger (Cambridge, England: Cambridge University Press, 1998), p. 99.

14. Trofim Desinovich Lysenko (1898–1976), the powerful director of genetic research in the Soviet Union during the Stalin era, rejected orthodox genetics, largely because it was thought to be a product of the "bourgeoisie" incompatible with dialectical materialism. He believed that wheat could be changed into rye when grown in appropriate conditions and other foolishness of the sort, wholly driven by ideological fervor.

15. "Lamarck," chap. 1 in Elie Faure, *Les Constructeurs: Lamarck-Michelet-Dostoievsky-Nietzsche-Cézanne* (Paris: Librairie Plon, 1950).

16. Claude Bernard, *Leçons sur les effets des substances toxiques et médicamenteuses* (Paris: Baillière et Fils, 1857).

17. The full text of Monod's Nobel lecture, as well as those of other Nobel laureates, may be consulted on the Internet, at www.nobelprize.org/medicine/laureates/index.html (visited May 25, 2005).

18. The statement is from the opening lines of Herbert Spencer's *First Principles* (London: Watts & Co., 1945). This work was first published in 1936, and by 1945 had gone through six editions.

19. Jacques Monod, *Le Hasard et la nécessité: essai sur la philosophie naturelle de la biologie moderne* (Paris: Seuil, 1970).

20. Giovanni Federspiel and Nicola Sicolo, "The Nature of Life in the History of Medical and Philosophic Thinking," *American Journal of Nephrology* 14 (1994): 337–343.

21. Michel Onfray, *Féeries anatomiques: généalogie du corps faustien* (Paris: Grasset & Fasquelle, 2003), pp. 106–112.

22. Spencer, *First Principles*, p. 54.

4. The Mystery of Procreation

1. G. R. Dunstan, ed., *The Human Embryo: Aristotle and the Arabic and European Tradition* (Exeter, England: Exeter University Press, 1990).

2. For a recent translation of Descartes's ideas on biology, see René Descartes, *The World and Other Writings,* trans. Stephen Gaukroger (Cambridge, England: Cambridge University Press, 1998).

3. William Harvey, *Disputations Touching the Generation of Animals,* trans., with introduction and notes, by Gweneth Whitteridge (Oxford, England: Blackwell Scientific Publications, 1981), p. 462.

4. *Soranus' Gynecology,* Book I, trans. Owsei Temkin with the assistance of Nicholson J. Eastman, Ludwing Edelstein, and Alan F. Guttmacher (Baltimore & London: Johns Hopkins University Press, paperbound edition, 1991), p. 12.

5. Quoted in Jean Rostand, *Esquisse d'une histoire de la biologie,* 5th ed. (Paris: Gallimard, 1945), p. 28.

6. Voltaire (François-Marie Arouet): "Dialogues d'Evhémère," in *Dialogues Philosophiques* (Paris: Garnier, 1966), pp. 442–447 (author's translation).

7. R.H.F. Hunter, *Physiology of the Graafian Follicle and Ovulation* (Cambridge, England: Cambridge University Press, 2003), pp. 1–20. See also G. Sarton, "The Discovery of the Mammalian Egg and the Foundation of Modern Embryology," *Isis* 16 (1931): 315–330.

8. William Cruikshank, "Experiments in Which, on the 3rd Day After Impregnation, the Ova of Rabbits Were Found in the Fallopian Tubes; and on the 4th Day of Impregnation in the Uterus Itself; with the First Appearances of the Foetus," *Philosophical Transactions of the Royal Society* 87 (1797): 197–214.

9. The classic works of these embryologists include Karl Ernst von Baer, *Uber Entwicklungsgeschichte der Thiere: Beobachtung und Reflexion* (2 vols.) (Königsberg: Gebrüdern Borntrager, 1828–1837). Much of this work is reproduced in English, albeit in a somewhat antiquated prose, in L. Y. Blayher, *History of Embryology in Russia from the Middle of the Eighteenth to the Middle of the Nineteenth Century,* trans. and ed. by H. I. Youseff and B. A. Maienschein (Washington, D.C.: Smithsonian Institution Libraries, 1982) (see chaps. 14–24). An additional source for important parts of von Baer's classics in English translation is J. M. Oppenheimer, *Autobiography of Karl Ernst von Baer* (Canton, Mass.: Science History Publications, 1986). The English version of many texts of difficult access by the nineteenth-century embryologists may be found in Howard B. Adelmann, *Marcello Malpighi and the Evolution of Embryology* (5 vols.) (Ithaca, N.Y.: Cornell University Press, 1966).

10. Hans Spemann and H. Mangold, "Über Induktion on Embryonalanlagen

durch Implantation artfremder Organisatoren," *Archiv für Mikroscopische Anatomie und Entwicklungsmechanik* 100 (1924): 599–638. In English, see Hans Spemann, *Embryonic Development and Induction* (New Haven, Conn.: Yale University Press, 1938).

11. The story of the first "test tube baby" was well told by the embryologist responsible for the first success of in vitro fertilization, in a book directed to the general public: Robert Edwards, *Life Before Birth: Reflections on the Embryo Debate* (New York: Basic Books, 1989).

12. Luc Montagnier, "Éloge de la réproduction naturelle," *Le Monde,* January 12, 2005.

13. Of the overwhelming number of books, monographs, and other publications devoted to the consequences of reproductive medicine technology, the following were found especially useful. The legal implications are made interesting by lively prose and a touch of humor in Lori B. Andrews, *The Clone Age: Adventures in the New World of Reproductive Technology* (New York: Owl Books, Henry Holt, 2000). Cloning is eloquently discussed by advocates from both sides, for and against, in Glenn McGee, ed., *The Human Cloning Debate* (Berkeley, Calif.: Berkeley Hills Books, 2000). A variety of perspectives, from the whimsical to the solemn, is found in Martha C. Nussbaum and Cass R. Sunstein, eds., *Clones and Clones: Facts and Fantasies About Human Cloning* (New York: W. W. Norton, 1999). A finely written criticism of some exaggerated claims by advocates of the new biotechnology is Richard Lewontin, *It Ain't Necessarily So: The Dream of the Human Genome and Other Illusions* (New York: New York Review of Books, 2000). Legal questions posed by prenatal life are discussed in Bonnie Seinbock, *Life Before Birth: The Moral and Legal Status of Embryos and Fetuses* (New York and Oxford, England: Oxford University Press, 1992). See also Anthony Dyson and John Harris, eds., *Experiments on Embryos* (London and New York: Routledge, 1990). For how the media, and in particular television, influence the perception on abortion, from a feminist standpoint, see Andrea L. Press and Elizabeth R. Cole, *Speaking on Abortion: Television and Authority in the Lives of Women* (Chicago: University of Chicago Press, 1999). What improvements in genetics biotechnology may mean for the individual and society is discussed thoroughly in Allen Buchanan, Dan W. Brock, Norman Daniels, and Saniel Wikler, *From Chance to Choice: Genetics and Justice* (Cambridge, England: Cambridge University Press, first paperback edition, 2001).

14. Ralph Jackson, *Doctors and Diseases in the Roman Empire* (London: British Museum Press, 2nd paperback edition, 1993), p. 86.

15. E. Ingerslev, "Rösslin's 'Rosengarten': Its Relation to the Past (the Muscio Manuscript and Soranus), Particularly with Regard to Podalic Version," *Journal of Obstetrics and Gynecology of the British Empire* 10 (1906): 297–325; see also vol. 12 (1907): 175–184; vol. 17 (1910): 3290–3332.

16. Adrian Wilson, *The Making of Man-Midwifery: Childbirth in England, 1660–1770* (Cambridge, Mass.: Harvard University Press, 1995), p. 5. See also J. W. Leavitt, *Brought to Bed: Childbearing in America, 1750–1950* (New York and Oxford, England: Oxford University Press, 1986).

17. Wilson, *The Making of Man-Midwifery,* pp. 65–67.

18. W. F. Bynum and Roy Porter, eds., *William Hunter and the 18th Century Medical World* (Cambridge, England: Cambridge University Press, reprint ed. 2000). See also C. H. Brock, ed., *William Hunter, 1718–1783: A Memoir by Samuel Foart Simmons and John Hunter* (Glasgow: University of Glasgow Press, 1983); and Sir Charles Illingworth, *The Story of William Hunter* (Edinburgh: E. and S. Livingstone, 1967).

19. William Hunter, *Reflections on Dividing the Symphysis Pubis. Supplement to Vaughan J. Observations on Hydrophobia* (London, 1778), cited in Brock, *William Hunter.*

20. For a learned discussion of the origin and history of the term "cesarean," see "Creative Etymology: 'Caesarean Section' from Pliny to Rousset," in Renate Blumenfeld-Kosinski, *Not of Woman Born: Representations of Caesarean Birth in Medieval and Renaissance Culture* (Ithaca, N.Y.: Cornell University Press, 1990), pp. 143–153. See also Dyre Trolle, *The History of Caesarean Section* (Copenhagen: C. A. Reitzel, 1982), p. 25.

21. Guy de Chauliac, *La grande chirurgie,* trans. E. Nicaise (Paris: Félix Alcan, 1890).

22. Ambroise Paré, *Œuvres complètes,* ed. J. F. Malgaigne (Paris: J. B. Baillière, 1840); see vol. 2, chap. 38, pp. 717–718.

23. Sherwin B. Nuland, *The Doctors' Plague: Germs, Childbed Fever, and the Strange Story of Ignác Semmelweis* (New York: W. W. Norton, 2003).

24. L. Yoles and S. Mashiach, "Increased Maternal Mortality in Cesarean Section as Compared to Vaginal Delivery: Time for Reevaluation," *American Journal of Obstetrics and Gynecology* 178 (1998, suppl.): S78.

25. C. S. Cotzias, S. Patterson-Brown, and N. M. Fisk, "Obstetricians Say Yes to Maternal Request for Elective Cesarean Section: A Survey of Current Opinion," *European Journal of Obstetrics, Gynecology, and Reproductive Biology* 97 (2001): 15–16. See also R. Gonen, A. Tamir, and S. Degani: "Obstetricians' Opinions Regarding Patient Choice in Cesarean Delivery," *Obstetrics and Gynecology* 99 (2002): 577–580.

26. J. Drife, "The Start of Life: A History of Obstetrics," *Postgraduate Medical Journal* 78 (2002): 311–315.

27. Jacalyn Duffin, "Women's Medicine and Medicine's Women: History of Obstetrics and Gynecology," chap. 11 in *History of Medicine: A Scandalously Short Introduction* (Toronto and Buffalo: University of Toronto Press, 1999, pp. 241–275).

28. Emily Wax, "In Rebuilding Sudan, Birth Often Brings Suffering and

Death," *The Washington Post*, Friday, March 4, 2005, p. A1. See also Kenneth•Hill, Carla Abou Zahr, and Teresa Wardlaw, "Estimates of Maternal Mortality for 1995," *Bulletin of the World Health Organization* 979, no. 3 (2001): 182–193.

29. Alexandre Minkowski, *L'Art de naître* (Paris: Odile Jacob, 1987), p. 221.

5. Pestilence and Mankind

1. Michel Drancourt and Didier Raoult, "Molecular Insights into the History of Plague," *Microbes and Infection* 4, no. 1 (January 2002): 105–109.

2. Carlo M. Cipolla, *Miasmas and Disease: Public Health and the Environment in the Pre-Industrial Age*, trans. Elizabeth Potter (New Haven, Conn.: Yale University Press, 1992).

3. *The Analects of Confucius: A Literal Translation with an Introduction and Notes by Chichung Huang* (Oxford, England: Oxford University Press, 1997), p. 81.

4. R. G. Cochrane and T. Frank Davey, *Leprosy in Theory and Practice*, 2nd ed. (Bristol, England: John Wiley and Sons, 1964), p. 374; O. K. Skinsnes, "Origin of Chaulmoogra Oil, Another Version," *International Journal of Leprosy* 40 (1972): 172–173.

5. An English translation of the best-known work of Julius Rosenbaum was published, although only a few copies of the book are extant; see Julius Rosenbaum, *The Plague of Lust: Being a History of Venereal Disease in Classical Antiquity* (Paris: G. J. Thieme for Charles Carrington, 1901).

6. Elizabeth Pennisi, "Genome Reveals Wiles and Weak Points of Syphilis," *Science* 281 (July 1998), 324–325.

7. William Osler, *The Principles and Practice of Medicine* (New York: D. Appleton, 1892), pp. 182–183.

8. Cynthia Landy, Phil Inouye, and David B. Hogan, "Veneral Disease and the Canadian Expeditionary Force in the First World War," *Annals of the Royal College of Physicians and Surgeons of Canada* 31, no. 8 (1998): 401–405.

9. Meyer Friedman and Gerald W. Friedland, *Medicine's Ten Greatest Discoveries* (New Haven, Conn.: Yale University Press, 1998).

10. The CDC report on venereal diseases can be read at www.cdc.gov/nchstp/dstd/Stats_trends2000.pdf (site visited February 11, 2006).

11. Al-Rhazi (Rhazes), *A Treatise on the Smallpox and Measles, by Abu Becr Mohammed ibn Zacariya ar-Razi (commonly called Rhazes)*, trans. from the original Arabic by W. A. Greenhill (London: Sydenham Society, 1847). See also *Medical Classics* 4, no. 1 (September 1939).

12. Robert Halsband, *The Life of Mary Wortley Montagu* (London: Oxford University Press, 1956); see also Robert Halsband, "New Light on Mary

Wortley Montagu's Contribution to Inoculation," *Journal of the History of Medicine and Allied Sciences* 8 (1953): 390–405.   •

13. Some biographies of Jenner are John Baron, *The Life of Edward Jenner* (2 vols.) (London: Henry Colburn, 1838); Edward F. Dolan, *Jenner and the Miracle of Vaccine* (New York: Dodd, Mead, 1960); W. R. Le Fanu, *A Bio-Bibliography of Edward Jenner (1749–1823)* (London: Harvey & Blythe, 1951).

14. In the English language, some biographies of Pasteur are Patrice Debré, *Louis Pasteur*, trans. Elborg Forster (Baltimore, Md.: Johns Hopkins University Press, 1998) (originally published in 1994 in French by Flammarion, on the occasion of the centenary of Pasteur's death; authored by a noted immunologist, this work lucidly develops many of the major scientific issues that engaged Pasteur); René Jules Dubos, *Louis Pasteur, Free Lance of Science* (New York: Da Capo Press, 1986) (reissue of the 1960 edition; this book, by a superb expositor of biomedical themes, emphasizes the philosophical aspects of Pasteur's investigations); René Jules Dubos and Thomas D. Brock, *Pasteur and Modern Science* (Washington, D.C.: American Society for Microbiology, 1998); René Vallery-Radot, *The Life of Pasteur* (New York: Doubleday, Page, 1916) (written by Pasteur's son-in-law; gives us a close look at the scientist; although currently out of print, other editions exist, including New York, Sun Dial Press, 1937); Gerald L. Geison, *The Private Science of Louis Pasteur* (Princeton, N.J.: Princeton University Press, 1995) (this book explores some of the aspects of Pasteur's work deemed questionable, such as his less-than-cautious rush to administer vaccines and the alleged incorporation of his rival's methods into his own, practices that in today's environment would elicit ethical concerns; reviewers have pointed out that this book, although clearly meritorious, unnecessarily emphasizes certain aspects of the savant's life that might appear negative from our vantage point; see review by M. F. Perutz: "The Pioneer Defended," *The New York Times Review of Books,* December 21, 1995).

15. Pasteur's first scientific article, originally in *Annales de Chimie et Physique,* vol. 24, was subsequently published as a monograph: Louis Pasteur, *Recherches sur les relations qui peuvent exister entre la forme cristalline, la composition chimique, et le sens de la polarisation rotatoire* (Paris: Bachelier, 1848).

16. Robert Koch, "Fortsetzung der Mittheilungen über ein Heilmittel genen Tuberkulose," *Deutsche Medizinische Wochenschrift* 17 (1891): 101–102; Robert Koch, "Weitere Mittheilung über das Tuberkulin," ibid., pp. 1189–1192.

17. Thomas McKeown, *The Role of Medicine: Dream, Mirage, or Nemesis?* 2nd ed. (Oxford: Blackwell, 1979).

18. N. P. Johnson and J. Mueller, "Updating the Account: Global Mortality of the 1918–1920 'Spanish' Influenza Pandemic," *Bulletin of the History of Medicine* 76 (2002): 105–115.

19. Rod Daniels, "In Search of an Enigma: 'The Spanish Lady,'" Mill Hill

1998 Essay of·the British National Institute for Medical Research (NIMT). See www.nimr.mrc.ac.uk/millhillessays/199/influenza1918.htm (visited February 18, 2006).

20. Michael T. Osterholm, "Preparing for the New Pandemic," *The New England Journal of Medicine* 352 (May 5, 2005): 1839–1842.

21. "Conspiracy Theories of HIV and AIDS," *The Lancet* 365, no. 9458 (February 5, 2005): 448.

22. The heartless Tuskegee deception has been the subject of a theater play, *Miss Evers' Boys* by David Feldshuh, which premiered in 1989 and has been performed several times since then. Books on the subject include James Jones, *Bad Blood: The Tuskegee Syphilis Experiment* (New York: Free Press, 1981 and 1992), and Susan Reverby, *Tuskegee Truths: Rethinking the Tuskegee Syphilis Study* (Chapel Hill: University of North Carolina Press, 2000). A short article, from the personal, sensitive viewpoint of a contemporary physician, is Joel D. Howell, "Trust and the Tuskegee Experiments," in Jacalyn Duffin, ed., *Clio in the Clinic: History in Medical Practice* (Oxford, England: Oxford University Press, 2005), pp. 213–225.

23. Edward Hooper, *The River: A Journey to the Source of HIV and AIDS* (Harmondsworth, England: Penguin, 1999).

24. Quoted by Kevin C. Kain in "Emerging Pathogens: The Birth of Plagues," *Annals of the Royal College of Physicians and Surgeons of Canada* 28, no. 3 (April 1995): 141–145. A comprehensive discussion of the relationship between climate change and infectious diseases was published by the World Health Organization (WHO): A. J. McMichael, D. H. Campbell-Lendrum, C. F. Corvalán, K. L. Ebi, A. Githeko, J. D. Scheragu, and A. Woodward, eds., *Climate Change and Human Health: Risks and Responses* (Geneva: WHO, 2003). See also Paul R. Epstein, "Climate Change and Human Health," *The New England Journal of Medicine* 353, no. 14 (2005): 1433–1436; and R. Sari Kovats and Andrew Haines, "Global Climate Change and Health: Recent Findings and Future Steps," *Canadian Medical Association Journal* 172, no. 4 (2005): 501–502.

25. D. H. Barouch, "Rational Design of Gene-Based Vaccines," *Journal of Pathology* 208 (2006): 299–319.

26. Joshua Lederberg, "Infectious History," *Science* 288 (April 14, 2000): 287–293.

6. CONCEPTS OF DISEASE

1. Mirko D. Grmek, *Diseases in the Ancient World*, trans. Mireille Muellner and Leonard Muellner (Baltimore: Johns Hopkins University Press, 1991); see Introduction.

2. Some of the medical articles on the beneficial effects of intestinal worms are J. V. Weinstock, R. Summers, and D. E. Elliott, "Helminths and Harmony," *Gut* 53, no. 1 (2004): 99–107; R. W. Summers, D. E. Elliott, and J. V. Weinstock, "Is There a Role for Helminths in the Therapy of Inflammatory Bowel Disease?," *National Clinical Practice of Gastroenterology and Hepatology* 2, no. 2 (2005): 62–63; R. W. Summers, D. E. Elliott, K. Qadir, J. F. Urban, and Thompson R. Weinstock, "*Trichiuris suis* Seems to Be Safe and Possibly Effective in the Treatment of Inflammatory Bowel Disease," *Inflammatory Bowel Disease* 11, no. 8 (2005): 783–784.

3. F. N. Silverman, "Introduction," in Daniel Bergsma, ed., *Skeletal Dysplasia* (New York: Stratton Intercontinental Medical Book Corp., 1974), pp. ix–xiv.

4. The most comprehensive translation (into French, with the Greek text on facing pages) of the Hippocratic Collection is Émile Littré, *Œuvres Complètes d'Hippocrate* (10 vols.) (Paris: Baillère, 1839–1861, reprinted 1961). Several volumes in Greek-English translation by W.H.S. Jones and E. T. Whitington are *Hippocrates* (Loeb Classical Library) (Cambridge, Mass.: Harvard University Press, 1923–1931, reprinted 1957–1959).

5. H.D.F. Kitto, *The Greeks* (Middlesex, England: Penguin, 1951), p. 188.

6. Quoted in Lynn Thorndike, *A History of Magic and Experimental Science During the First Thirteen Centuries of Our Era*, vol. 1 (New York: Columbia University Press, 1993), p. 728.

7. Walter Pagel, *The Smiling Spleen: Paracelsianism in Storm and Stress* (Basel: Karger, 1984); Pagel, *Paracelsus: An Introduction to Philosophical Medicine in the Era of the Renaissance*, rev. 2nd ed. (Basel: Karger, 1982).

8. William Harvey, *The Anatomical Exercises: De Motu Cordis and De Circulatione Sanguinis in English Translation*, ed. Geoffrey Keynes (New York: Dover Publications, 1995), p. 91. This work is a reprint of an anonymous translation of the seventeenth-century classic. Its merit is that it preserves the language of that century and thus conveys the impression that Harvey himself is speaking.

9. Ibid.

10. L. S. King, "Empiricism and Rationalism in the Works of Thomas Sydenham," *Bulletin of the History of Medicine* 44 (1970): 1–11. Sydenham's works have been collected in Thomas Sydenham, *The Works of Thomas Sydenham, M.D. Translated from the Latin edition of Dr. Greenhill. With a life of the author by R. G. Latham* (2 vols.) (London: The Sydenham Society, 1848–1850).

11. René-Théophile-Hyacinthe Laënnec, *Traité de l'auscultation médiate et des poumons et du coeur*, 4th ed. (Paris: J. S. Chaudé, 1837). For an English translation, see W. Hale-White, *Translation of Selected Passages from de l'Auscultation Médiate* (preceded by a 24-page biography of Laënnec)

(New York: Wood, 1923). For another short biography of Laënnec, see G. B. Webb, "René Théophile Hyacinthe Laënnec," *Annals of Internal Medicine* 9 (1927): 27–59.

12. E. H. Ackerknecht, "Broussais or a Forgotten Medical Revolution," *Bulletin of the History of Medicine* 27 (1953): 323–343.

13. Ibid.

14. A short biography of Rudolf Virchow is contained in an anonymous obituary in the September 13, 1902 issue of *British Medical Journal*, pp. 795–802. See also E. H. Ackerknecht, *Rudolf Virchow: Doctor, Statesman, Anthropologist* (Madison: University of Wisconsin Press, 1953).

15. Rudolf Virchow, *Die Cellularpathologie* (Berlin: A. Hirschwald, 1858). The second edition of this classic work has an English translation by Dr. Frank Chance, approved by Virchow himself: *Cellular Pathology as Based Upon Physiological and Pathological Histology* (New York: Dover Publications, 1971). It contains an important introductory essay by L. J. Rather entitled "The Place of Virchow's Cellular Pathology in Medical Thought."

16. Oswei Temkin, "The Scientific Approach to Disease: Specific Entity and Individual Sickness," chap. 30 in *The Double Face of Janus and Other Essays in the History of Medicine* (Baltimore: Johns Hopkins University Press, 1977), pp. 441–455.

17. K. Faber, *Nosography: The Evolution of Clinical Medicine in Modern Times* (New York: Paul B. Hoeber, 1930), pp. 207–208.

18. Stephen J. Kunitz, "Classifications in Medicine," chap. 9 in Russell C. Maulitz and Diana E. Long, eds., *Grand Rounds* (Philadelphia: University of Pennsylvania Press, 1988), p. 293.

19. Ruy Pérez-Tamayo, *El concepto de enfermedad: su evolución a través de la historia* (2 vols.) (Mexico City: Fondo de Cultura Económica, 1988), vol. 2, p. 96.

7. The Diagnostic Process

1. Janet D. Howell, *Technology in the Hospital: Transforming Patient Care in the Early Twentieth Century* (Baltimore, Md.: Johns Hopkins University Press, 1995), p. 136.

2. Bettyann Holtzmann Kevles, *Naked to the Bone: Medical Imaging in the Twentieth Century* (New Brunswick, N.J.: Rutgers University Press, 1997). This is a well-researched, highly readable account of the major advances in medical imaging and their impact upon art, society, and the view we form of ourselves.

3. Ibid., p. 27.

4. Bruno Haliqua, *Histoire de la Médecine,* 2nd ed. (Paris: Masson, 2004), p. 24.

5. *The Yellow Emperor's Classic of Internal Medicine,* trans., with an introductory study, by Ilza Veith (Berkeley: University of California Press, 1972, reprinted from 1949 edition).

6. The quoted author is Thomas Baker, a Fellow of St. John's at Cambridge, in a work entitled *Reflections upon Learning, Wherein Is Shewn the Insufficiency Thereof, in Its Several Particulars: In Order to Evince the Usefulness and Necessity of Revelation. By a Gentleman* (London: Bosville, 1700; London: Knapton and Wilkin, 1714, 1727), quoted in Lu Gwei-Djen and Joseph Needham, *Celestial Lancets: A History and Rationale of Acupuncture and Moxa* (New York: Routledge Curzon, 2002), p. 37.

7. R. J. Bush, "Urine Is a Harlot or a Lier," *The Journal of the American Medical Association* 208 (1969): 131–134; M. H. Haber, "Pisse Prophecy: A Brief History of Urinalysis," *Clinical Laboratory Medicine* 3 (September 8, 1988): 415–430; Ruth Harvey, "The Judgement of Urines," *Canadian Medical Association Journal* 159 (1988): 1482–1484; William I. White, "A New Look at the Role of Urinalysis in the History of Diagnostic Medicine," *Clinical Chemistry* 37, no. 1 (1991): 119–125.

8. L. C. MacKinney, *Early Medieval Medicine* (Baltimore: Johns Hopkins University Press, 1937), pp. 46–48.

9. S. Chen, L. Zieve, and V. Mahadevan, "Mercaptans and Dimethyl Sulfide in the Breath of Patients with Cirrhosis of the Liver," *Journal of Laboratory and Clinical Medicine* 75 (1970): 628–635.

10. Quoted in François Millepierres, *La Vie quotidienne des médecins au temps de Molière* (Paris: Hachette, 1965), p. 265 (author's translation).

11. Guido Majno and Isabelle Joris, "The Microscope in the History of Pathology," *Virchow's Archives of Pathology: (A) Pathologic Anatomy* 360 (1973): 273–286.

12. Guido Majno, *The Healing Hand: Man and Wound in the Ancient World* (Cambridge, Mass.: Harvard University Press, 1975), pp. 150–153.

13. F. González-Crussi, *On Seeing: Things Seen, Unseen and Obscene* (New York: Overlook Press, 2006).

14. Antonio Benivieni, *De abditis nonnullis ac mirandis morborum et sanationum causis,* trans. Charles Singer, with a biographical appreciation by Esomond R. Long (Springfield, Ill.: Charles C. Thomas, 1954). A facsimile reprint of the Latin text, with English text on opposite pages.

15. The first edition of the *Sepulchretum* was printed in Geneva by Leonard Chouet, publisher, in 1679. Little is known about Bonet's life. See E. E. Irons, "Théophile Bonet (1620–1689). His Influence on the Science and Practice Of Medicine," *Bulletin of the History of Medicine* 12 (1942): 623–665.

16. The enunciation of the cellular theory was recorded in a German publication, of which there is an English-language translation: Theodor Schwann,

Microscopical Researches into the Accordance in the Structure and Growth of Animals and Plants, trans. Henry Smith (London: Sydenham Society, 1847). The most important part of this work, namely the formulation of the cellular theory, is on pages 186–215 and may be consulted online at www.72.14.203.104/search?q=cache:FjBbtprXESMJ:mechanism.ucsd.edu/ ~bill/teaching/philbio/THEODOR%2520SCHWANN.htm+theodor+ schwann+microscopical+researches&hl=en&gl=us&ct=clnk&cd=5 (visited April 20, 2006).

8. THERAPY

1. Normal Taylor, "From Shen Nung to Brigham Young," in *Plant Drugs That Changed the World* (New York: Dodd, Mead, 1965).
2. Article published in *Le Monde* online, written by a special envoy to Yaoundé, Africa (author's name not given): "Une Plante chinoise contre le paludisme," November 25, 2005.
3. Max Neuburger, *Essays in the History of Medicine* (New York: Medical Life Press, 1930), p. 42.
4. Motolinia (pen name of Fr. Toribio de Benavente), *Memoriales o libro de las cosas de la Nueva España y de los naturales de ella,* ed. E. O'Gorman (Mexico City: U.N.A.M., 1971) (originally published 1541), p. 160. See also Fr. Juan de Torquemada, *Monarquía Indiana* (7 vols.), ed. Miguel León Portilla (Mexico City: U.N.A.M., 1975–1983) (originally published 1615); see vol. 3, p. 325.
5. Bernard R. Ortiz de Montellano, *Aztec Medicine, Health, and Nutrition* (New Brunswick, N.J.: Rutgers University Press, 1990), pp. 181–188.
6. Uriel García Cáceres, *Juan del Valle y Caviedes: cronista de la medicina. Historia de la medicina en el Perú en la segunda mitad del siglo XVII* (Lima, Peru: Universidad Peruana Cayetano Heredia, 1999), pp. 42–48.
7. La Fontaine, *Œuvres Complètes* (2 vols.), 2nd edition. See vol. 2: *Œuvres diverses: Poème du Quinquina, à Madame la Duchesse de Bouillon* (Paris: Gallimard, Collection La Pléiade, 1958), p. 67.
8. Pliny, *Natural History,* 2nd ed., Loeb Classical Library, Book XXVI, 131–134, trans. W.H.S. Jones (Cambridge, Mass.: Harvard University Press, 1992) (first published 1956), p. 365.
9. Ibid., Book XXV, iii, pp. 6–8 (vol. 8 of Loeb Classical Library), pp. 139–141.
10. M. Bariety and Ch. Coury, *Histoire de la médecine* (Paris: Fayard, 1963), p. 320.
11. Herodotus, *The Histories,* Book II, pp. 75–77. See Aubrey de Sélincourt's translation, with notes by A. R. Burn (New York: Penguin Books, 1972), p. 158.
12. Ibid., Book II, p. 83, p. 160.

13. Bruno Haliqua and Bernard Ziskind, *Medicine in the Days of the Pharaohs*, trans. M. B. DeBevoise, with a foreword by Sonald B. Redford (Cambridge, England: The Belknap Press of Harvard University Press), pp. 13–14. See also L. Viso and J. Uriach, "The Guardians of the Anus and Their Practice," *International Journal of Colorectal Diseases* 10 (1995): 229–231.

14. François Millepierres, *La Vie quotidienne des médecins au temps de Molière* (Paris: Hachette, 1965), p. 144.

15. Julius Friedenwald and Samuel Morrison, "The History of Enema with Some Notes on Related Procedures," *Bulletin of the History of Medicine* 8 (1940): 68–114.

16. Regnier de Graaf, *De Clysteribus* (1668), translated from Latin into French by an anonymous translator under the title *L'Instrument de Molière* (Paris: Damascène, Morgand & Charles Fatout, 1878), pp. 67–68 (author's translation).

17. R. Lunarotti (article in Italian), "An Old and Peculiar Resuscitation Procedure: Clyster of Tobacco Smoke," *Acta Anaesthesiologica* 19, no. 4 (July–August 1968): 657–663.

18. Erwin H. Ackerknecht, *A Short History of Medicine*, rev. ed. (Baltimore, Md.: Johns Hopkins University Press, 1982), p. 205.

19. Roy Porter, *The Greatest Benefit to Mankind: A Medical History of Humanity* (New York: W. W. Norton, 1997), p. 512. Henceforth referred to as *The Greatest Benefit to Mankind*.

20. Ibid., p. 513.

21. Ibid., p. 520.

22. Edgar I. Irving, "Origins and Development of Shock Therapy in Psychiatry," in *The Origins of the Healing Art* (New York: Philosophical Library, 1978), pp. 83–190.

23. Richard Hunter and Ida Macalpine, *Three Hundred Years of Psychiatry (1535–1860)* (New York: Oxford University Press, 1963), p. 534.

24. Jean-François Reverzy, *L'Homme et sa folie*, in *Histoire des moeurs*, vol. 3: *Thèmes et systèmes culturels* (Paris: Gallimard, Encyclopédie de la Pléiade, 1991), pp. 767–802.

9. Some Concluding Thoughts

1. William Edward Hartpole Lecky, *History of European Morals*, vol. 1 (London: Longmans, Green, 1902) (first printed 1869), p. 158.

2. Roy Porter, *The Greatest Benefit to Mankind*, p. 715.

3. U.S. Bureau of the Census, Historical Statistics I:55, Department of Commerce, may be consulted online at www.infoplease.com/ipa/A0005140.html; Richard Cooper, Robert Cohen, and Abbas Amiry, "Is the

Period of Rapidly Declining Adult Mortality Coming to an End?," *American Journal of Public Health* 73 (1983): 1091–1093.

4. Sherr L. Murphy, "Deaths: Final Data for 1998," *National Vital Statistics Reports* 48 (July 2000).

5. "Industry's Role in Hypertension" (editorial), *The New York Times,* May 30, 2006; see www.nytimes.com/2006/05/30/opinion30tue3.html?th&emc=th.

6. Mircea Eliade, "The Occult and the Modern World," chap. 4 in *Occultism, Witchcraft and Cultural Fashions: Essays in Comparative Religions* (Chicago: University of Chicago Press, 1976), pp. 47–68.

7. See, e.g., the descriptions of acupuncture and moxibustion by the explorer E. Kaempfer (1651–1716), reprinted as Engelbert Kaempfer: *Exotic Pleasures: Fascicle III, Curious Scientific Medical Observations,* translated, with a commentary, by Robert Carruba (Carbondale: Southern Illinois University Press, 1996), pp. 108–140.

Index

F. GONZÁLEZ-CRUSSI was born in Mexico, obtained his M.D. degree from the National University of Mexico in 1961, and pursued postgraduate training in Canada and the United States, becoming a U.S. citizen in 1973. He is currently professor emeritus of pathology at Northwestern University Feinberg School of Medicine in Chicago and was head of laboratories at Chicago's Children's Memorial Hospital until his retirement in 2001. Besides having written over two hundred articles in journals of his medical specialty, he has distinguished himself as a literary essayist. His endeavor to fuse science and literature, applying his biomedical background to humanistic and philosophical topics, has earned him high praise from critics at home and abroad. He has authored fifteen books (five in his native Spanish), among which are *Notes of an Anatomist, Suspended Animation: Six Essays on the Preservation of Bodily Parts, The Day of the Dead: And Other Mortal Reflections, There Is a World Elsewhere, The Five Senses, On Being Born: And Other Difficulties,* and *On Seeing: Things Seen, Unseen, and Obscene.* He lives in Chicago with his wife, Dr. Wei Hsueh, also a retired pathologist and a biomedical researcher.

A NOTE ON THE TYPE

The principal text of this Modern Library edition
was set in a digitized version of Janson, a typeface that
dates from about 1690 and was cut by Nicholas Kis,
a Hungarian working in Amsterdam. The original matrices have
survived and are held by the Stempel foundry in Germany.
Hermann Zapf redesigned some of the weights and sizes for
Stempel, basing his revisions on the original design.

Printed in the United States
by Baker & Taylor Publisher Services

Printed in the United States
by Baker & Taylor Publisher Services